**Aurélio Ferreira Borges**
**Mauri C. Mazutti**
**Maria Anjos C. S. Borges**

# Food security in the Western Amazon:

**Aurélio Ferreira Borges**
**Mauri C. Mazutti**
**Maria Anjos C. S. Borges**

# Food security in the Western Amazon:

## A study applied to project development

**ScienciaScripts**

**Imprint**



Cover image: www.ingimage.com

This book is a translation from the original published under ISBN 978-613-9-66019-3.

Publisher:
Sciencia Scripts
is a trademark of
Dodo Books Indian Ocean Ltd. and OmniScriptum S.R.L publishing group

120 High Road, East Finchley, London, N2 9ED, United Kingdom
Str. Armeneasca 28/1, office 1, Chisinau MD-2012, Republic of Moldova, Europe
Printed at: see last page
**ISBN: 978-620-7-98273-8**

# Summary

## Presentation

This book, in the form of chapters, is part of a broader process of reflection on the elaboration of projects that have been launched to obtain resources for scholarships, capital and funding from national and international development agencies. As an important turning point in this process of systematising and producing knowledge, it aims to refine the initial findings of this ongoing debate on contemporary society and its relationship with science and culture, responding to specific urgent demands for a grounding in the practice of preparing projects today.

The choice of chapters was guided by a number of observations. The first was that contemporary societies have been undergoing transformations in all the various dimensions of the production of human existence; the second was that these changes have materialised in a specific way in Brazilian society, due to the way Brazil has been inserted into the world throughout its history and also due to the economic, political, social and cultural relations that take place at national level; and the third was that these changes taking place in the world and in the country, which redefine ideals, ideas and social practices, have repercussions on the processes of project development, leading to a redefinition of the content and way of structuring technical-scientific training, since in contemporary societies, science has been responding in a specific way to the need to valorise capital, to the ethical-political conformation of bourgeois sociability and also to the effective popular demand for access to socially produced knowledge.The ideas presented here by the authors thus fulfil the function of subsidising the discussions that have been taking place in the elaboration of projects related to the contemporary philosophical and socio-historical guidelines of their practice of science and research historically committed to the training of Brazilian workers, within the frameworks of their existence.

The authors.

Happy reading!

# CHAPTER 1

## Description of the object to be carried out up to information on the technical and managerial capacity of the project proponent

Mauri Carlos Mazutti[1]

Aurelio Ferreira Borges[2]

Maria dos Anjos Cunha Silva Borges[3]

Edmar da Costa Alves[4]

Diego Soares Carvalho[5]

## Summary

With the slogan Real Food in the Countryside and in the City: for rights and food sovereignty, the project aims to highlight the socio-cultural dimensions of food and nutrition security in order to bring food production and consumption closer together; to build bridges between urban and rural areas; to value agro-biodiversity and organic production, fresh and regional foods, respect for black and indigenous ancestry, Africanity and the traditions of all traditional peoples and communities, and the recovery of the identities, memories and food cultures of the population of Rondônia's Territories of Citizenship. The food system, which tends towards commodity production, is marked by the hegemony of large-scale monoculture production with high mechanisation, which has massified the use of pesticides and transgenics, alongside market control by foreign corporations.

## 1. Description of the object to be carried out

Table 1 - Identification of the proposal.

| | |
|---|---|
| 1. *project title* | Performance of forests, agroforestry and agroextractivist systems for agroecological and organic food production in the Western Amazon. |
| 2. *Project coordinator* | Aurelio Ferreira Borges |
| 3. *Executing institution* | Federal Institute of Education of Rondônia |

[1] Bachelor of Laws. Volunteer researcher on projects funded by the National Council for Scientific and Technological Development (CNPq).

[2] PhD in Forestry Engineering. Researcher and coordinator of projects funded by CNPq. Full Professor of the Master's Programme in Professional and Technological Education at the Federal Institute of Education of Rondônia (IFRO).

[3] Master in Literature. Volunteer researcher on CNPq projects.

[4] Master in Food Engineering. Server at IFRO.

[5] PhD in Biology. Professor at IFRO.

| | |
|---|---|
| *4. Collaborators* | University of Aberdeen (UK), University of Paris III, National University of Colombia, UFRRJ, UNIR, Facimed University of Cacoal. |
| *5. Support mode* | Maintenance of the Centre for Studies in Agroecology and Organic Production (NEA) (Approved by CNPq, process 463352/2014-9, call 11/2014, beginning Jan./2015) and ending 31.07.*2017,* entitled Extension and research centre for sustainable rural development in the Vale do Guaporé-RO Territory. |
| *6. Line* | Line 2: Maintenance of the Centre for Studies in Agroecology and Organic Production (NEA). |
| *7. General objective* | Improve agroecology and organic production through scientific research and technological extension, with rural productive inclusion in specific population groups, with an emphasis on traditional peoples and communities and other vulnerable social groups in the western Amazon. |

The research, extension and transfer of the technologies generated will initially be carried out in the Territories, the Vale do Guaporé-RO Rural Identity Territory (TRVG) and the Rio Machado-RO Rural Identity Territory (TRRM), located in the Western Amazon (Figures 1 and 2). The population of these two (2) Territories is 257,527 inhabitants.

The project will be institutionalised for at least another 3 years at the Federal Institute of Education of Rondônia (IFRO) and the Federal University of Rondônia (UNIR), via a call for tenders. Institutionalisation will involve the inclusion of capital expenditure, funding and grants. Institutionalisation will include the other five (5) Rural and Citizenship Territories in Rondônia.

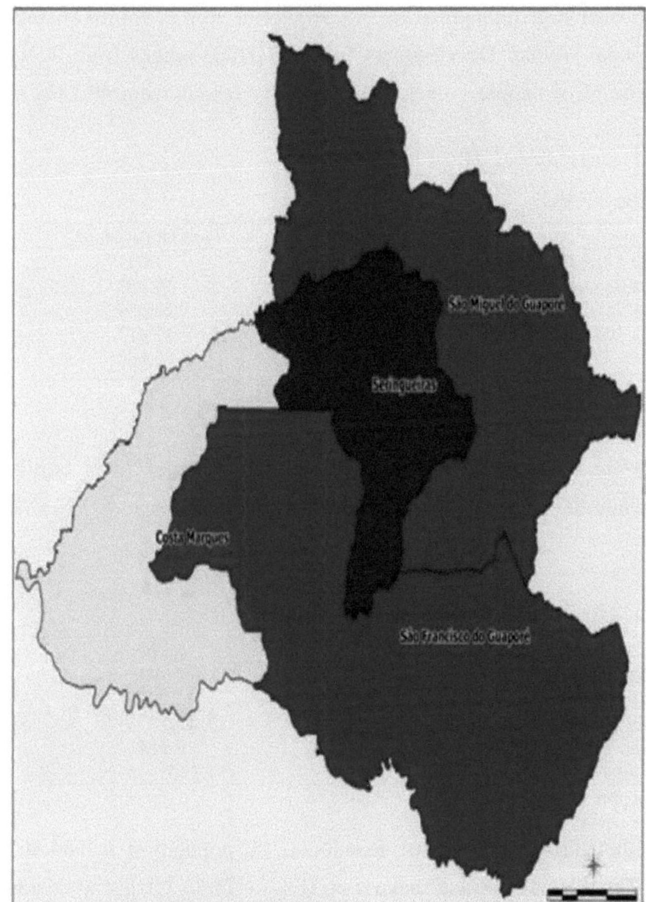

Figure 1 - Municipalities of the TRVG.

Source: http://sit.mda.gov.br/mapa.php.

The research, extension and transfer of technologies generated will also be used to include them in the TRVG's Territorial Plan for Sustainable Rural Development and Solidarity (PTDRSS), which is currently being drawn up, and to update the TRRM's PTDRSS. Actions developed under the Territories of Citizenship Programme (PTC) will be taken up, with the Territorial Development Committees (CODETER) acting as focal points for the implementation of the programme.

Plan.

The TRVG is located in the northern region of the state of Rondônia. Its population is 73,060 inhabitants spread over four municipalities. Its territorial area is approximately 27,180.68 $Km^2$ . The Municipal Human Development Index (MHDI) ranges from 0.598 (low) to 0.646 (medium). The % of people vulnerable to poverty ranges from 49.18% to 75.42% (Table 2).

Table 2 - Municipalities in the TRVG.

| *Municipality* | *GDP per capita* | *MHDI* | *% Vulnerable to* | *Population people* |
|---|---|---|---|---|
| Costa Marques | 10.571,09 | 0,611 | 75,42 | 17.031 |
| São Francisco do | 14.636,61 | 0,611 | 71,91 | 19.353 |
| São Miguel do Guaporé | 18.253,88 | 0,646 | 49,18 | 24.059 |
| Rubber trees | 13.507,83 | 0,598 | 63,02 | 12.617 |
| *Total population* | | → | | 73.060 |

Source 1: http://cidades.ibge.gov.br.
Source 2: Source: http://sit.mda.gov.br/mapa.php.

In the TRVG there are 2,993 settled agrarian reform families and 7,484 family farming establishments. The number of people employed in family farming is 23,671 (Table 3).

Table 3 - Variables and values for sociodemographic data from the TRVG.

| *Variable* | *Value* |
|---|---|
| No. of Settled Families - Agrarian Reform | 2.993 |
| Number of Projects - Agrarian Reform | 17 |
| Reformed Area - Agrarian Reform (in hectares) | 314.517 |
| No. of family farming establishments | 7.484 |
| Staff employed in family farming | 23.671 |

Source: http://sit.mda.gov.br/mapa.php.

The TRRM is in the northern region of Rondônia. Its population is 184,467 inhabitants spread over the 7 municipalities that make it up (Figure 2). Its territorial area is 19,045.86 $Km^2$ . The MHDI index ranges from 0.617 (low) to 0.718 9 (medium). The % of people vulnerable to poverty ranges from 51.18% to 79.86% (Table 4).

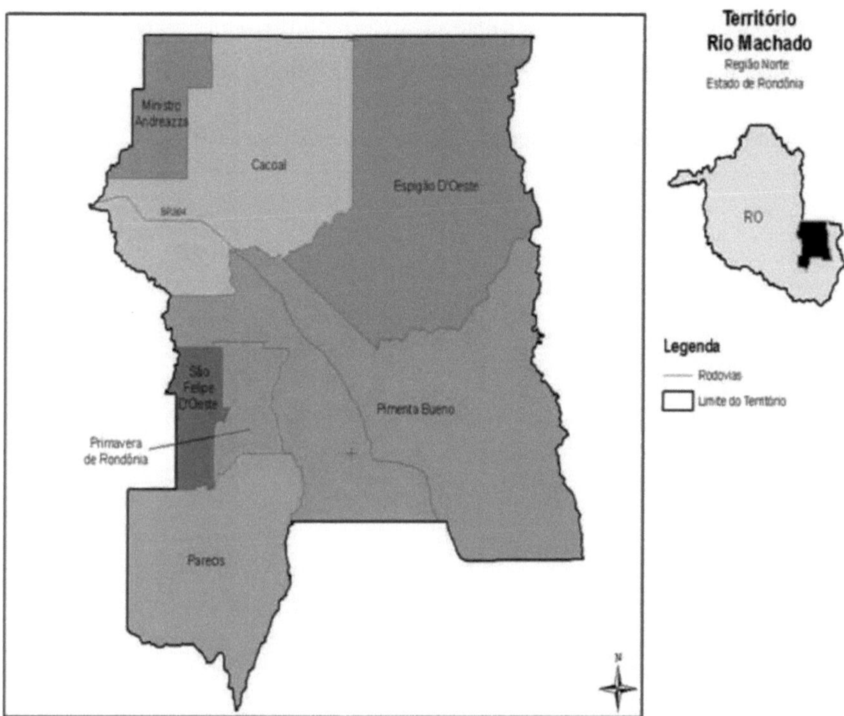


Figure 2 - TRRM municipalities.

Source 1: http://cidades.ibge.gov.br.

Source 2: Source: http://sit.mda.gov.br/mapa.php.

The TRRM has seven municipalities (Table 4).

Table 4 - TRRM municipalities and their social indicators.

| Municipality | GDP per capita R$ | MHDI | % Vulnerable to *poverty* | *Inhabitants* |
|---|---|---|---|---|
| Cacoal | 19.593,03 | 0,718 | 51,28 | 87.877 |
| Espigão do Oeste | 15.550,85 | 0,672 | 57,62 | 32.712 |
| Minister Andreazza | 16.080,42 | 0,701 | 61,40 | 10.786 |
| Parecis | 14.449,59 | 0,617 | 51,74 | 5.802 |
| Pimenta Bueno | 22.896,99 | 0,754 | 79,86 | 37.786 |
| Spring | 14.912,14 | 0,691 | 59,30 | 3.456 |
| São Felipe do Oeste | 11.701,94 | 0,694 | 68,04 | 6.048 |
| *Total population* | | → | | *184.467* |

Source: http://cidades.ibge.gov.br.

In the TRRM there are 1,940 settled agrarian reform families and 9,636 family farming establishments. There are 30,872 people in family farming (Table 5).

Table 5 - Variables and values for sociodemographic data from the TRRM.

| *Variable* | *Value* |
|---|---|
| No. of Settled Families - Agrarian Reform | 1.940 |
| Number of Projects - Agrarian Reform | 18 |
| Reformed Area - Agrarian Reform (in hectares) | 557.309 |
| No. of family farming establishments | 9.636 |
| Staff employed in family farming | 30.872 |

Source: http://sit.mda.gov.br/mapa.php.

Below are the associations of traditional peoples and communities and family farmers that exist in these two (2) Territories.

## 1.1. Trvg

The following types of social organisation are fully active in this Territory:

- Municipality of São Miguel do Guaporé-RO

*THE* Rural Producers' Association of Linha Noventa.
*THE* Association of Rural Producers of the AGRIFU Line.
*THE* São Miguel Rural Association for Mutual Aid (ARSAPAM).
. Quilombola Community of Jesus Association (AQCJ).
*THE* Association of Fairgoers (ASFESMIG).

*T* Uru-Eu-WAu-WAu indigenous land.
*T* Rio Branco Indigenous Land.

-Municipality of Seringueiras

. Boa Esperança Rural Producers Association (ASPRUBE).
. Seringueiras Farmers' Association (ASERIPAM).
. Bom Futuro Rural Family Producers Association (ASPROBOM).
. OLAVO PIRES Rural Producers Association (ASPROP).
. Association of United Rural Producers (ASPRONI).
. Vale do Guaporé Rural Producers' Co-operative (COOPERVAGS).

. Uru-Eu-WAu-WAu indigenous land.
*T* Rio Branco Indigenous Land.

-Municipality of São Francisco do Guaporé

. São Francisco do Guaporé Fishermen's Colony;
. Association of Rural Women of the District of Porto Murtinho (ASMUPORT).
. Association of Rural Producers União e Força (ASPRUF).
. Associação Remanescentes Quilombolas de Pedras Negras do Guaporé (AQPNEG).
. Quilombola Remnants Association of Santo Antônio do Guaporé.
. Uru-Eu-WAu-WAu Indigenous Land Association.

**-Municipality** of Costa Marques

. Associação dos Quilombolas do Forte Príncipe da Beira (ASQUIPFORT).
. Association of Quilombolas of Santa Fé (ASQUISANTAFÉ).
. Costa Marques-RO Fishermen's Colony Z4;
. Ouro Fina Rural Producers Association Line 12 (ASPROFINO)
. Association of Rural Producers in the Bom Jesus Sector (ASPROBOM).
. Association of Rural Producers of the Primavera Sector (ASPROP).
. Massaco Indigenous Land Association.
. Association of Extractivists of the Rio Calcário Extractive Reserve.

## 1.2. Trrm

The following types of social organisation are fully active in this Territory:

*THE* Beekeepers' Association (ACA).
*THE* Surui Indigenous Community Association.
. Mutual Aid Association (COOPERAZZA, ACOMAC).
*THE* Agricultural and Agroindustry Electricity Association (ENEAGRO).
*THE* Association of Farmers' Groups (AGAOV).
. Women's Association (ASMUQ).
*THE* Association of Rural Mining Producers (ASPRORAM).
. Rural Producers' Associations (APREVEC, GUARAPARI).
. Association of Olive Growers (AOCVC, ASPROCHOP).
*THE* Association of Alternative Family Producers (APROFAMA).
. Agricultural Workers Family Association (ARFATA).

*THE* Small Producers Association (APROLU).
*THE* Association of Agricultural Settlement Producers (ACOMAF, APROJE).
. Chacareiros Association (ARCA).
. Mixed Agriculture Co-operative (COOPEMAGRI).
. Management Committees of the Zero Hunger Programme in the Municipalities.
. Community of Pomeranian Culture.
. Municipal Councils (health, rural development, education, food security).
. Credit Cooperative (CREDICACOAL, CREDITAG).
. Federation of Agricultural Workers of Rondônia (FETAGRO).
*THE* Roosevelt Indigenous Land Association, of the Cinta Larga indigenous people of Espigão do Oeste-RO.

## 2. Axes, specific objectives, strategies, targets and initiatives

### 2.1. Production

Specific objective 1: To expand the production, handling and processing of organic and agro-ecological products in the TRRM and TRVG, with the following aims

priority public family farmers, agrarian reform settlers, traditional peoples and communities and their economic organisations, micro and small rural enterprises, cooperatives and associations, also considering urban and peri-urban agriculture.

- Strategies:

*o* Guide the preparation of new Infrastructure and Services Projects for the Territories (PROINF)
based on the principles of agroecology and organic production, making them a priority.
. Ensure that at least 30 per cent of the technical team in training and capacity-building activities is made up of women.
. Promoting women's agroecological transition and production through the Programme of Rural Women's Productive Organisation, reviewing its guidelines, forms and instruments of organisation, management bodies and social participation.

Table 6 - Target 1: Enable access to financing policies for family farmers, agrarian reform settlers, traditional peoples and communities, urban and peri-urban agriculture, organic and agro-ecological producers.

| *Initiative* | *Partner institutions* | *Indicator* | *Physical execution target* | | |
|---|---|---|---|---|---|
| | | | *2017* | *2018* | *9019* |
| Train at least 200 ATER technicians, financial agents, members of rural workers' unions and movements, forestry workers, indigenous leaders, riverside dwellers, quilombolas, gypsy leaders, about the credit lines related to the sociobiodiversity. | MAP FUNAI | Empowered people | - | 200 | - |
| To register and train 400 family farmers, members of rural unions, forestry, indigenous, riverine, quilombola, gypsy and other traditional peoples, Family Production Units, to be included in the National Register of Organic Products. | MAP FUNAI | N° of traditional peoples TRVG/TRRM in the National Register of Organic Products | - | 200 | 200 |

Source: http://www.mda.gov.br/planapo/.

## 2.2 Use and conservation of natural resources

Specific objective 2: Promote, expand and consolidate processes of access, sustainable use, management, recomposition and conservation of natural resources and ecosystems in general in the TRVG and TRRM Territories.

- Strategies:

. Promote territorial networks aimed at the rescue, conservation *in situ, on farm and* free use of creole, local and traditional varieties.

. To recognise and value traditional practices and knowledge associated with the use and management of medicinal and aromatic plants and herbs by women.

*G* Guarantee and strengthen the participation of rural youth in natural resource management and conservation processes.

Table 7 - Target 2. Promote processes of access, sustainable use, management, recomposition and conservation of natural resources and ecosystems.

| *Initiative* | *Partner institutions* | *Indicator* | *Physical execution target* | | |
|---|---|---|---|---|---|
| | | | *2017* | *2018* | *9019* |
| Promote the propagation of seedlings and varietal and creole seeds to at least 300 traditional peoples, with trials to evaluate genetic resources and products from socio-biodiversity and agro-ecology, in the Territorial Seeds and Seedlings Programme, with the participation of the | MDA | Female family farmers | - | 100 | 200 |

| | | | | | |
|---|---|---|---|---|---|
| To support the structuring of community seed banks and Seed Processing Units (UBS) in the interest of agroecology and organic production, encouraging gender equality in the management of the banks. | EMBRAPA | Structured seed banks | | 1 | |
| Implementing the "Creole and traditional agroecological and organic seed bank" experiment, *NDVI, solar heater, photovoltaic panel for isolated communities, mini rainwater harvesting cistern, organic vegetable production and socio-environmental mapping. | EMBRAPA | Experiments carried out | - | 1 | 1 |

**Note: *_Normalised Difference Vegetation Index,_ used to determine land cover in a GIS (Geographic Information System). For this purpose, satellite images from the TM Landsat-5, ETM+ Landsat-7 CCD CBERS-2 and CCD and HRC CBERS 2B sensors are used. The remote sensing and geographic information system data are integrated with models for predicting**
**Soil losses have been characterised as important tools in assessing the environment, such as mapping and quantifying areas of agroecological production of fruit, vegetables and legumes on indigenous lands, in family farming. Also the question of water availability, by mapping springs, estimating land cover, estimating biomass and forecasting crops from agroecological and organic production.**

Source: http://www.mda.gov.br/planapo/.

## 2.3. Knowledge

Specific objective 3: To expand the capacity for building and socialising knowledge on agroecology and organic production systems, by valuing local culture and exchanging knowledge and internalising the agroecological perspective in teaching, research and extension institutions and environments in the TRVG and TRRM territories.

- Strategies:

*D* Providing agro-ecological and sustainable systems-based TSA, valuing the role of women and young people, with a territorial focus.
*I* Intensify the systematisation of academic and scientific production and agroecological knowledge and the provision of appropriate technical teaching material for technicians, farmers, producers and students.

Table 8 - Target 3. Promote education with an agroecological and organic production systems approach for students, family farmers, extractivists, fishermen, agrarian reform settlers, indigenous peoples, traditional peoples and communities, rural youth and women.

| *Initiative* | *Partner institutions* | *Indicator* | *Physical execution target* | | |
|---|---|---|---|---|---|
| | | | *2017* | *2018* | *1019* |
| Organise 1 agroecology experience for 400 | INCRA | Realised | - | 1 | - |

| ATER technicians, students, family producers, extractivists and fishermen, agrarian reform settlements, indigenous peoples and traditional communities. | | experience | | | |
|---|---|---|---|---|---|
| Hold the 1st Seminar on Education in Agroecology and Organic Production, in partnership with the Agroecology Association. | MDA | Seminar held | - | - | 1 |

Source: http://www.mda.gov.br/planapo/.

## 2.4 Commercialisation and consumption

Specific objective 4: To improve the marketing of organic, agroecological and socio-biodiversity-based products on local, regional, national and international markets and in public procurement, and to increase the consumption of organic, agroecological and socio-biodiversity-based products.

- Strategies:

R Carry out permanent actions to publicise the production and consumption of organic and agro-ecological food, with campaigns involving non-governmental and governmental partners from the different federal states.

. Improve and stimulate government purchases from producers of traditional peoples and communities and farmers converting to organic production and agroecological transition.

Table 9 - Target 4. Promote the commercialisation and consumption of agroecological and socio-biodiversity products.

| *Initiative* | *Partner institutions* | *Indicator* | *Physical execution target* | | |
|---|---|---|---|---|---|
| | | | *2017* | *2018* | *9019* |
| Hold events with 40 public managers responsible for PNAE and PAA purchases, encouraging the purchase of food from organic, agroecological and socio-biodiversity sources in school feeding menus. | MAP | No. of events PNAE/PAA week | 1 | 1 | - |
| To train 200 vocational education and university students, family farmers, producers in agro-ecological and organic transition, higher education and federal institute teachers, rural youth and ATER agents in eco-gastronomy through a partnership with the *slow food* movement (http://www.slowfoodbrasil .com/). | SLOW FOOD | Nº of people trained | - | 100 | 10 |

Source: http://www.mda.gov.br/planapo/.

## 2.5. Land and Territory

Specific objective 5: To guarantee access to land and territories as a way of promoting the ethno-development of traditional peoples and communities, indigenous peoples and agrarian reform settlers.

- Strategies:

*A* Increasing the participation of associations/cooperatives of indigenous peoples and traditional communities in management and marketing support programmes such as the Quilombola Seal of Brazil.

. Support ethno-development projects and actions.

Table 10 - Target 5. Make land reform settlements, sustainable use conservation units, territories of traditional peoples and communities and indigenous peoples priority areas for promoting agroecological production.

| *Initiative* | *Partner institutions* | *Indicator* | *Physical execution target* | | |
|---|---|---|---|---|---|
| | | | *2017* | *2018* | *9019* |
| Increase the participation of the 50 associations of indigenous peoples and traditional communities in the TRRM and TRVG territories in the Mais Gestão Programme*. | MDA | Number of institutions served | 1 | 1 | 1 |
| To publicise the Quilombos do Brasil Seal among the quilombola associations in the TRRM and TRVG territories and to advise the associations on how to access the Seal. | SEPPIR*, Palmares Foundation | No. of publicity actions | 1 | 1 | - |
| To publicise the Indigenous Seal of Brazil to indigenous people and advise them on access to the seal, in order to qualify traditional indigenous production and access institutional and private markets. | FUNAI | Nº dissemination actions | 1 | - | 1 |

Note: * Mais Gestão promotes the inclusion of administration techniques in associations of people from traditional communities, agrarian reform settlers and family farming by qualifying management systems (organisation, production and marketing). The aim is to guarantee access to the markets of the National School Feeding Programme (PNAE). It is a programme that uses ATER methodology for family farming cooperatives.
* SEPPIR = Special Secretariat for Policies to Promote Racial Equality, of the Ministry of Justice and Citizenship (http://www.seppir.gov.br/).

Source: http://sit.mda.gov.br/mapa.php.

## 2.6. Sociobiodiversity

Specific objective 6: To promote the recognition of socio-cultural identity, the improvement of social organisation and the guarantee of the rights of indigenous peoples, traditional peoples and communities and family farmers.

- Strategies:

*o* To advise associations on negotiations regarding access to genetic heritage and traditional knowledge associated with biodiversity.

. Adjust forestry and socio-biodiversity product regulations, consolidating family community forest management and the sustainable management and use of native species.

. Increasing the presence of socio-biodiversity products and foods in the PAA, PNAE and other institutional purchasing processes.

Table 11 - Target 6. Expand the inclusion of socio-biodiversity products in institutional markets and local and regional markets.

| *Initiative* | *Partner institutions* | *Indicator* | *Physical execution target* | | |
|---|---|---|---|---|---|
| | | | *2017* | *2018* | *9019* |
| Lobbying purchasing bodies to include socio-biodiversity foods in institutional purchasing processes (*PAA and *PNAE), through the State Technical Marketing Chambers. | CONAB | Training No. of Experiences | - | 1 | 1 |
| Define proposals for priority areas for structuring 6 Local Productive Arrangements (APLs) for socio-biodiversity, focusing on the following chains: açaí, rubber, babassu, Brazil nuts, pequi and pirarucu. | EMBRAPA ICMBIO CONAB FUNAI | Training No. of Experiences | - | - | 1 |
| Hold 3 training workshops for indigenous peoples, traditional peoples and family farmers to access commercialisation policies agroecological and organic, for gender equity and youth. | MMA CONAB | Workshops held | 1 | 1 | - |

Note: *PAA = the federal government's Food Acquisition Programme; *PNAE = the National School Feeding Programme.

Source: http://sit.mda.gov.br/mapa.php.

## 3. Estimated time for implementation of 24 months

Table 12 - Summary for the project timetable.

| *Activity* | *Location* | *Implementation target (dates)* | | |
|---|---|---|---|---|
| | | *2017* | *2018* | *9019* |
| Organic horticulturist FIC course. | IFRO | - | 20.08.18 - 28.11.18 | - |
| Lecture Acquisition of PNAE and PAA products. | TRVG TRRM | 03.10.1 7 | 07.03.18 | |
| Lecture 0 seal indigenous peoples and quilombolas of Brazil, to the indigenous communities of the TRVG/TRRM. | S. Rural Espigão D'Oeste | 23.11.1 7 | 13.03.18 | 13.03.19 |
| Workshop on policies to support the commercialisation of agro-ecological and organic production. | S. Rural S. M. Guaporé | 02.12.1 7 | 02.04.18 | 02.03.19 |
| Seminar on Agroecology and organic production. | IFRO | | 01.10.18 04.10.18 | |
| Experiment A: Mother bank of food, oilseed, fibre, fodder and green manure seeds. | Andreazza | | 04.02 a 11.02.18 | 04.04 a 11.05.19 |

| Experiment B: Silvicultural agroforestry model | Kilo. Santa Fe | 04.10 a 11.12.1 7 | 04.03 a 11.04.18 | - |
|---|---|---|---|---|
| Experiment C: Normalised Difference Vegetation Index (NDVI) with GIS (Geographic Information System). | TRVG TRRM | 04.11 a 31.12.1 7 | | |
| Experiment D: Photovoltaic panel for isolated communities. | TRVG | | 04.05 a 11.06.18 | |
| Field day | TRVG TRRM | - | - | 01.08-19 |
| Publication of a scientific article | Magazine | | | 01.0830.08 .19 |
| Publication of a book chapter | Publisher | | | 01.0830.08 .19 |
| Video publication | Centre | - | 01.08 30.08.18 | 01.0830.08 .19 |

Source: http://sit.mda.gov.br/mapa.php.

# 4. Identification of the team and description of their professional profile

The researchers taking part in the research, their institution and their role in the proposed research are presented. As participants in the research, we should consider the current existence of two doctoral students with FAPERO/CAPES scholarships and two (2) undergraduate students, who will still be selected according to their academic records and interviews. There are also two (2) researchers from a UNASUR country (Colombia), a researcher from UFRRJ, professors from IFRO, a professor from the University of Paris III and a professor from the University of Aberdeen (Scotland) (Table 13).

Table 13 - Project team.

| *Name* | *Team function* | *Agroecology/organic production experiences* |
|---|---|---|
| *Aurelio Ferreira Borges* | Transferring technologies, Coordinating | Skills in Food and Nutrition Security (FNS), agroecology and organic production, sustainable rural development, popular education. |
| *Marcos Aurélio Anequine de Macêdo* | Transfer technology, research | Family Farming; Agro-biodiversity and agro-extractivism; Project Evaluation; Solidarity Economy; Rural Extension; Participatory Methodologies. |
| *Creuci Maria Caetano* | Transfer technology, research | Professor of the Doctorate in Agroecology at the Universidad N. Colombia (UNC). Phylogenetic resources. Plant Genetics. |
| *Edilberto Syryczyk* | Transfer technology, research | Ethnoscience, ethnomathematics, indigenous culture, Nhambiquara ethnology, agroecology, organic production. |
| *Valdemir Lúcio Durigon* | Transfer technology, research | Coordinator of the Agroecology Technician programme at UFRRJ. Agronomy: Soil Management and Conservation and Agroecology, remote sensing |
| *Maria dos Anjos* CNPq Fellow. | Transferring technologies, research | Territorial Advisor on Gender; Organic Agriculture; Associativism; Culture and Education in the Amazon; Territorial Development. |

Note: *NEDET^t* **Territorial Development Extension Centre. These are projects** supported by CNPq, public call n. 11/2014. As a result of the MDA/SPM-PR/CNPq cooperation, the NEDETs are responsible for providing technical support, advice and monitoring for the work of the Territorial Committees, bringing academia closer to the implementation of public policies through university research and extension activities. They also seek to involve various aspects of social management processes and the implementation of public policies aimed at the social and economic development of the territory.

Source: http://sit.mda.gov.br/mapa.php.

## References

AGROECOLOGICAL BRAZIL. **National Plan for Agroecology and Organic Production - Planapo**: 2016-2019 / Interministerial Chamber for Agroecology and Organic Production. - Brasília, DF: Ministry of Agrarian Development, 2016. 89 p.

CITIES, I. B. G. E. data on the Internet. **Rio de Janeiro: IBGE**. Available at:< http://www. ibge. gov. br>. Accessed on 29 July 2018.

MINISTRY OF AGRARIAN DEVELOPMENT (MDA). **Agroecological Brazil: national plan for agroecology and organic production - planapo 2016 to 2019**. Available at:<http://www.mda.gov.br/planapo>. Accessed on: 29 July 2018.

MINISTRY OF AGRARIAN DEVELOPMENT (MDA). **Territorial notebooks**. Available at:<http://sit.mda.gov.br/download.php>. Accessed on: 29 July 2018.

# CHAPTER 2

## Justification up to expected project results

Aurelio Ferreira Borges

Mauri Carlos Mazutti

Maria dos Anjos Cunha Silva Borges

Edmar da Costa Alves

Edilberto Fernandes Syryczyk[6]

Diego Soares Carvalho

## Summary

The productivist logic of maximum profit generates socio-environmental impacts that are expressed in deforestation, in compromising agro-biodiversity and organic production, which affects traditional peoples and communities in the western Brazilian Amazon. The agribusiness model oppresses the realisation of the human right to adequate food. One of the relevance and impacts of this proposal is to set up a task force to characterise, register and improve rural extension, scientific research and the availability of technologies in agroecology and organic production for families from specific population groups, the so-called GPTEs, consisting of indigenous families, quilombola families, gypsy families, families belonging to terreiro communities, extractivists, riverine communities, artisanal fishermen, family farmers, agrarian reform settlers, campers, beneficiaries of the National Land Credit Programme and those affected by infrastructure projects.

## 1. Justification

The guidelines of the National Policy for Agroecology and Organic Production (PNAPO), which are also those that guide the preparation and implementation of the Planapo, are as follows, according to Article 3 of Decree No. 7,794/2012 (BRASIL, 2012):

I - Promoting food and nutrition sovereignty and security and the human right to adequate and healthy food, by offering organic and agro-ecological products free from

[6] PhD in Science and Maths Education. Full Professor of the Masters in Professional and Technological Education at IFRO.

contaminants that endanger health. II - Promoting the sustainable use of natural resources, observing the provisions that regulate labour relations and favour the well-being of owners and workers.

III - conservation of natural ecosystems and restoration of modified ecosystems, by means of agricultural production and forestry extraction systems based on renewable resources, with the adoption of cultural, biological and mechanical methods and practices that reduce polluting waste and dependence on external inputs.

IV - Promoting fair and sustainable food production, distribution and consumption systems that improve the economic, social and environmental functions of agriculture and forest extraction and prioritise institutional support for the beneficiaries of Law No. 11.326.

V - Valuing agro-biodiversity and socio-biodiversity products and encouraging local experiments in the use and conservation of plant and animal genetic resources, especially those involving the management of local, traditional breeds and varieties.

VI - Increasing the participation of rural youth in organic and agroecological production.

VII - contributing to the reduction of gender inequalities through actions and programmes that promote women's economic autonomy.

The internalisation of these guidelines requires a high degree of articulation between Pnapo and other public policy agendas. Pnapo aims to articulate actions to support the production, consumption and marketing of agroecological and organic food, as well as related programmes in the areas of knowledge, culture and environmental preservation. In this sense, it works together to build and strengthen policies such as technical assistance and rural extension, food security, agrarian reform, climate change, among others, as well as being strengthened by structuring and advancing these agendas.

The Centre for Study, Research and Extension in Agroecology, Organic Production and Sustainability for the Western Amazon was created with the aim of developing educational, research and extension actions aimed at improving the agroecological transition and organic production for traditional peoples and communities in social vulnerability in the Rural and Citizenship Territories: Vale do Guaporé-RO Rural Territory (TRVG) and Rio Machado-RO Rural and Citizenship Territory (TRRM).

Located at IFRO's Cacoal-RO campus, the Agroecology and Organic Production Centre emerged from an extension and research project approved by CNPq/MDA call number 11/2014 (CNPq process no. 463352/2014-9). 463352/2014-9), which through strategies enabled the institutional consolidation of the Extension and Research Centres (NEDETs) of these two (2) Territories as an academic space, in the development of research and teaching and extension activities aimed at training traditional peoples and communities in social vulnerability (families from specific population groups, the so-called GPTEs, indigenous families, women of the waters and forests, rural youth, quilombolas, gypsy families, families belonging to terreiro communities, extractivists, riverine communities, artisanal fishermen, family farmers, agrarian reform settlers, campers, beneficiaries of the National Land Credit Programme, affected by infrastructure projects). The NEDETs receive funding from the CNPq to carry out actions that enrich the process of emancipation of GEPTE peoples and sustainable agriculture within the Territories of Citizenship.

This NEDET has a technical course in Agroecology, which works with agricultural and extractive production systems based on agroecological principles and organic production system techniques. It develops integrated actions, combining the preservation and conservation of natural resources with the social and economic sustainability of production systems. It works to conserve soil and water. Helps with integrated family farming actions, taking into account the sustainability of smallholdings and production systems. Participates in actions to conserve and store raw materials and industrialise agro-ecological products. Works in public, private and third sector institutions, agroecological certification institutions, research and extension institutions, parks and nature reserves.

a) Target audience: this is aimed at those who have completed primary school or are currently studying at another public secondary technical vocational education institution and wish to transfer to IFRO.

b) Course load: 4,200 hours (3,502 clock hours of 50 minutes).

c) Duration of the course: A minimum of 3 (three) and a maximum of 6 (six) years. d) Timetable: Full-time.

The set of subjects that make up the Professionalising Core of the course is made up of the subjects summarised below (Table 1).

Table 1 - Summary of subjects on the technical course in agroecology and production at IFRO, Cacoal campus (TRRM and TRVG).

| *Discipline* | *Workload* |
|---|---|
| Introduction to Agroecology | 80 |
| Agroecological soil management | 120 |
| Agroecological Plant Management I | 120 |
| Agroecological Plant Management II | 120 |
| Agroecological Animal Management I | 80 |
| Orientation to Research and Professional Practice | 40 |
| Rural Property Management | 80 |
| Entrepreneurship | 80 |
| Legislation and Certification of Agroecological Products | 80 |
| Biodiversity and Conservation in Agroforestry Systems | 120 |
| Water Resources Management | 120 |
| Family Farming and Rural Development | 120 |

Source: http://portal.ifro.edu.br/nossos-cursos?id=257.

More information about IFRO's Technical Course in Agroecology can be found at the following link: http://portal.ifro.edu.br/nossos-cursos?id=257

Conventional farming has been one of the factors responsible for the degradation of the Amazonian environment. This is jeopardising the satisfaction of the needs of future generations. Therefore, avoiding degradation has become urgent, and it is under this premise that the adoption of agroecological and organic production systems in Rural and Citizenship Territories is an option for achieving efficiency in the use of natural resources, which are already scarce, since a large part of national agricultural production belongs to the category of GPTE groups.

In this way, the NEDETs seek to build a collective consciousness for Sustainable Rural Development, carrying out actions within the TRVG, the TRRM and the surrounding area. Students and GPTEs have won several victories in the fight for projects and resources that take into account the activities of a group of vulnerable populations and a differentiated social movement.

## 2. Audience of the proposal

a) Students from the professional and technological education network and universities.

b) Family farmers, under the terms of the Family Farming Law, Law no.

11.326/2006 (BRASIL, 2006).

c) Producers in agroecological transition or involved in agroecological or organic production.

d) Teachers from Higher Education Institutions, institutions of the Federal Network of Professional, Scientific and Technological Education and State Public Institutions of Professional and Technological Education.

e) Technical Assistance and Rural Extension Agents (ATER).

f) Workers, including family farmers, foresters, aquaculturists, extractivists and fishermen.

g) Beneficiaries and dependents of federal cash transfer programmes, among others, who meet the criteria specified under the Brazil without Poverty Plan.

h) People with physical disabilities.

i) Indigenous peoples, quilombola communities and other traditional communities.

j) Priority audiences of federal government programmes that are associated with the Training Grant.

k) Adolescents and young people under socio-educational measures.

l) Students who have completed their secondary education in public schools or in private institutions on a full scholarship.

Research, extension and transfer of the technologies generated and available will initially be carried out in the Vale do Guaporé-RO Rural Territory (TRVG) and the Rio Machado-RO Citizenship Territory (TRRM), located in the Western Amazon of Brazil (Figures 1 and 2).

The project will then be institutionalised for at least another 3 years at IFRO and the Federal University of Rondônia (UNIR), to continue the Agroecology and Organic Production plan after the 24-month duration of this proposal. The model used to manage the project will be the MS Project Gantt chart (VARGAS, 2016). A Gantt chart transforms task names, start date, completion date and duration into horizontal chart bars. You can quickly visualise the order in which the tasks should take place and which tasks are dependent on each other.

The Gantt chart is effective when managing a complex project. This model will be reconstructed in the project as proposed by KERZNER (2006) and BOUER (2005). The authors propose an assessment scheme for each maturity level of the Project Management Maturity Model (PMMM). For level 1, Common Language, an 80-question questionnaire is applied covering the knowledge areas of the PMBoK (Project Management Body of Knowledge), which provides a reference structure for project management covering nine knowledge areas: (i) integration; (ii) scope; (iii) time; (iv) cost; (v) quality; (vi) human resources, (vii) communication; (viii) risk; (ix) procurement. The results of the questionnaire allow the organisation to obtain a picture of its level of maturity in terms of the common language for project management (Figure 1). For level 2, Common Processes, of the PMMM model, the project management life cycle will be identified, which can be broken down into five phases: embryonic, acceptance by senior management, acceptance by management, growth and maturity. KERZNER (2006) emphasises that the growth phase is critical, as it marks the beginning of the creation of a project management process, and warns that standardisation should be sought in methodologies for planning, executing and controlling projects. At level 3 of the PMMM model, Singular Methodology, the author proposes a questionnaire protocol to assess six characteristics of the so-called hexagon of excellence: integrated processes, culture, leadership support, training and education. Level 4 of the PMMM model, Benchmarking, seeks to assess the extent to which an organisation uses the process and practices characteristic of benchmarking to improve its project management. For level 5 of the PMMM maturity, called Continuous Improvement. There are practices adopted by the organisation to improve and disseminate the learning accumulated through the implementation of project management in the organisation (Figure 1).

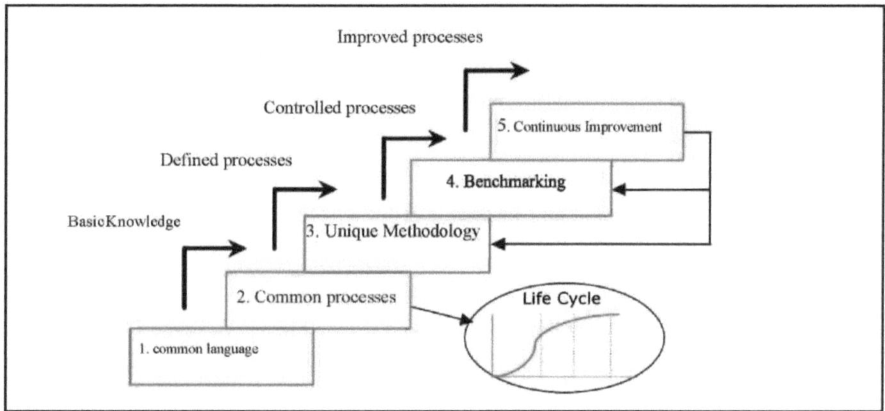

Figure 1 - Project Management Maturity Model (PMMM).

Source: Kerzner (2006); Bouer (2005).

## 3. Characterisation of the problem to be solved

The agri-food system is undergoing transformations in ways of living, living, communicating, cooking and eating that do not reflect the rich, diverse and lively dynamics of society. The traditional Brazilian menu, represented by rice, beans, manioc, corn, pumpkin, fruit, vegetables and legumes, is being threatened by the advertising appeal of industrialised and ready-to-eat products, with an excess of sodium, sugars, fats, preservatives, pesticides and GMOs. In urban areas, eating out has become an imposition, making it difficult to cope with the impacts of these changes. Restrictions on healthy food and the means of production, such as land and water, markets, and food prices are the main inflationary factor in Brazil for meals outside the home. The food system, with its tendencies towards commodity production, is marked by the hegemony of large-scale monoculture production with high mechanisation, which has massified the use of pesticides and transgenics, alongside market control by foreign corporations.

The productivist logic of maximum profit generates socio-environmental impacts that are expressed in deforestation, in compromising agro-biodiversity, which affects traditional peoples and communities. The agribusiness model oppresses the realisation of the human right to adequate food (VAZ, 2018). A major difficulty for sustainable rural development lies in the lack of dialogue between scientific and popular knowledge, since these can be incongruous. Farmers and traditional peoples display specific, contextualised knowledge, which often differs from academic knowledge that is focused on generalisations and epistemological-methodological issues of its own, which can be difficult to align with farmers' localised and pragmatic knowledge. In addition, there are difficulties due to the current training of researchers, cultural profile, structure of institutions and the dispute over fields of knowledge (PRATES JÚNIOR et al., 2016).

From the authors' perspective, the local knowledge of rural populations and peoples should be valued for what they see as legitimate and powerful constructions in their domains and valid according to the cognitive and epistemic criteria of their own cultural patterns. The rapprochement between scientific and non-scientific fields is capable of expanding the sense of what is possible, through the emergence of competing explanations that open up space for a relativist perspective, even if it is not of Feyrabend's "anything goes" type (GOLL et al., 2018).

The difficulties in strengthening agroecology and organic production systems at local/regional level stem from the difficulty in distinguishing between cognitive and social values in science, since social values can play a legitimate role in the adoption of a strategy (or methodological rules) or in the application of scientific knowledge, but when it comes to the acceptance of theories, only cognitive values and empirical data should play a fundamental role. For this reason, LACEY (2006) reinforces the importance of three epistemic virtues required by the scientific community: a) objectivity: the quest to produce the most reliable knowledge, emphasised above in terms of cognitive criteria; b) neutrality: values and ethical judgements cannot be inferred from scientific results, i.e. the naturalistic fallacy is not allowed; c) autonomy: scientific institutions must not be burdened with extra-scientific interests, especially those related to the hegemonic powers linked to capital and the market that exacerbate social inequalities and compress cultural diversity.

According to PRATES JÚNIOR et al. (2016), there is a need for theoretical generalisations, explanations and predictability within the realm of science so that not just case studies or singular statements are established. Therefore, Agroecology and Organic Production in search of balance in the field of scientific research also opens up space for context-specific research that can transform practices into pedagogical activities, such as action research, and not just for the research itself, which they still dominate. Compartmentalisation, which is sometimes important in order to delve deeper into certain issues, weakens the technical-scientific system.

Alternatively, agroecology and organic production need to establish a kind of "orchestration", in which the different approaches are associated and incompatibilities are used in the search for solutions, but without reductionism or the utopian search for the unification of the sciences. It is a question of resisting excessive specialisation, in an invitation to take steps outside our specialities, into new research possibilities, so that, as stated, it is possible to visualise, study, understand and manage the complexity of the agri-food system

(PRATES JÚNIOR et al., 2016).

The authors also add that in order to be widely characterised as a field of scientific mastery, scholars of Agroecology and Organic Production need to establish attempts to improve their body of theory, in constant competition and replacement of incomplete issues or those less accepted by the scientific community, even from the perspective of a paradigmatic shift. And despite recognising exchanges with other areas of knowledge, it is not possible to establish syntheses between the fields involved, as they are distinct and to a certain extent incongruous (MAYR, 2008).

Agroecology and organic production, in turn, have established dialogues with non-scientific knowledge more frequently than other scientific fields. The challenge is to gain legitimacy through the values of current science, in order to gain ground and break with the premises that condition the dominant scientific model. In order to be characterised as a scientific field, agroecology scholars need to improve their body of theory, constantly competing with and replacing issues that are less accepted by the scientific community (MAYR, 2008).

According to BORGES et al. (2014), the occupation of Brazil's territorial space provides sample patterns that the literature defines as "frontiers". This agricultural frontier is characterised by the displacement of territorial transformations, based on the insertion of new industrial and rural production technologies.

From the authors' perspective, the Brazilian territory was taken over from the coast to the interior, and this method of occupation "into the hinterland" was called frontier expansion, characterised by different "fronts". This method of occupation was characterised by pragmatism and immediacy, where the preservation of the original vegetation was always relegated to second place. In Rondônia, farming appeared as the foundation of this process of installation of the Pioneer Front, during the initial occupation of the territory, and the Expansion Front, referring to the capitalist use of the land, especially from the beginning of the 70s of the 20th century, with the growing intensification of occupation, as well as a rising capitalist assimilation of the land.

According to SILVA (2008), the state of Rondônia is characterised by the concentration of large areas reserved for preservation through Conservation Units (UC) and Indigenous Lands (TI), which make up a total area of 90,301.6 $Km^2$ , representing 38% of the state's total area. The Conservation Units are distributed over a total of 54, including one Environmental Protection Area, five Ecological Stations, eleven State Sustainable Yield Forests, three National Forests, three State Parks, two National Parks, four Biological Reserves and 25 Extractive Reserves, totalling 53,153 km2, 22.4% of the state. Of this total, 41 are state-owned and 13 federal, covering 21,681 and 31,472 square kilometres, respectively. According to the SNUC (National System of Conservation Units) classification, 40 are for sustainable use and 14 for full protection, totalling 24,384 and 28,769 Km2 respectively. The Indigenous Lands are distributed over 28 areas totalling 51,415 km2. There are three areas of overlap between indigenous lands and conservation units, demonstrating a land problem that needs to be addressed with greater attention.

## 4. Expected results

With the slogan "Real Food in the Countryside and in the City: for rights and food sovereignty", this proposal aims to highlight the socio-cultural dimensions of food and nutritional security in order to bring food production and consumption closer together; to build bridges between the urban and the rural; to value agro-biodiversity and organic production, in natura and regional foods, respect for black and indigenous ancestry, Africanity and the traditions of all traditional peoples and communities, the rescue of the identities, memories and food cultures of the population of the seven (7) Citizenship Territories of Rondônia, Brazil.

In Brazil, a wide range of programmes are aimed at family-based food production and rural populations, such as the Food Acquisition Programme (PAA), the National Programme for Strengthening Family Farming (Pronaf), the National Policy for Technical Assistance and Rural Extension (Pnater), the Brazil Quilombola Programme and the National Plan for the Sustainable Development of Traditional Peoples and Communities.

Food insecurity still persists among indigenous peoples, traditional peoples and communities, urban populations living on the streets and in poverty, people with disabilities, segments of the black population and, notably, women breadwinners. The proposal seeks to eradicate institutional racism and overcome manifestations of prejudice of all kinds, especially against people in vulnerable situations (GOLL et al., 2018).

The food system in the TRVG and TRRM Territories, in line with global trends in commodity production, is marked by the hegemony of the large-scale patronage production model of monocultures with high mechanisation, which has massified the use of pesticides and transgenics, alongside growing market control by large foreign corporations. As well as enshrining the historical concentration of land ownership in Brazil, the productionist logic of maximum profit generates serious socio-environmental impacts, which are expressed in deforestation, the compromise of biodiversity and agro-biodiversity, as well as threats to the rights won in the 1988 Constitution (BRAZIL, 2017), such as the Proposed Constitutional Amendment (PEC) 215/00, which affects indigenous peoples and quilombolas. The agribusiness model, as well as major infrastructure projects, oppress the realisation of the human right to adequate food.

The TRVG and TRRM Territories are made up of thousands of peasants, family farmers, indigenous peoples, quilombola communities, traditional peoples of African descent, terreiro peoples and other traditional peoples and communities who produce the healthy food that makes up our diet, account for the largest share of jobs in rural areas, and whose production and organisational systems are better suited to sustainable and diversified production, even though they occupy a much smaller total area than agribusiness and are under permanent pressure from it.

Women in the cities, countryside, waters and forests of the TRVG and TRRM territories play a strategic role in guaranteeing food and nutrition security, as they are food producers. However, gender inequalities persist: even today, the vast majority of women work triple shifts and black women suffer double discrimination on the grounds of gender and race. Tackling these and other injustices is one of the main tasks of this project, as it seeks to take action against racism and sexism, as well as to implement specific education and training policies that denaturalise the sexual division of labour and gender violence, in order to make it possible to build new paradigms of shared responsibilities between men and women in the public and private spheres. The aim is to ensure that these women are recognised as political subjects in the process of building rural development, as well as their leading role in guaranteeing food and nutrition sovereignty and security.

The aim is to improve the inductive and regulatory role of the Brazilian state in the spheres of food production, supply, distribution, commercialisation and consumption in the TRVG and TRRM. Models will be provided for regulatory actions that can control the expansion of monocultures and the actions of transnational corporations; proposals that maintain the moratorium on the use of terminator seeds (purposely sterile plants), seeking the precautionary principle in the control of the release and commercialisation of transgenics; proposals will be drawn up for areas free of transgenics and agrochemicals; proposals will be indicated to regulate the labelling of these products, advertising and other market practices, especially with a view to protecting children. The aim is to collaborate in the readjustment of health legislation for food of animal origin and for artisanal, traditional and family production, as well as to develop taxation and regulation mechanisms for ultra-processed, high-salt, high-sugar, high-fat, transgenic and bi-fortified product industries (GOLL et al., 2018).

Development policies are proposed for socio-biodiversity, agro-ecology and organic production, with measures for access to local markets and the means of production in these Territories, to natural goods and seeds, as well as the incorporation of agro-ecological principles, methods and social technologies and the guarantee of the rights of family farmers, indigenous peoples, quilombola communities, traditional peoples of African origin, and other traditional peoples and communities to the free use of agro-biodiversity. Among the various measures, guaranteeing the expansion of Creole seed banks throughout the state of Rondônia and the recognition and dissemination of traditional knowledge associated with biodiversity.

The aim is to improve, qualify and expand programmes such as the National Food Acquisition Programme (PAA) and the National School Feeding Programme (PNAE), with differentiated per capita amounts for indigenous people and quilombolas, public purchases, Technical Assistance and Rural Extension (ATER) actions and the National Programme for Strengthening Family Farming (Pronaf). The National Plan for Agroecology and Organic Production needs to be strengthened and the National Programme for Reducing the Use of Pesticides (Pronara) put into effect, as a way of stimulating the agroecological transition process, expanding and popularising the supply of healthy food. The aim is to improve policies for reducing the use of pesticides and to propose technologies for monitoring the rate of pesticide contamination.

Urgent action must be taken on the availability of and access to water, given the threats to the realisation of this right. Climate change, resulting from the ultra-capitalist exploitation of nature, is causing droughts and floods in different places. This context has affected water consumption and local production systems, with direct impacts on the population's food and nutritional security and on the socio-economic development of many regions. Actions such as the participatory and politicised construction of cisterns, as well as the adoption of social strategies and technologies aimed at guaranteeing water for human consumption, basic health services, education and the production of healthy food can reduce these impacts ((GOLL et al., 2018).

The project's activities will contribute to sustainable rural development in its environmental, social and economic aspects, enabling the beneficiary public to build and socialise knowledge and technologies related to Agroecology, Organic Production Systems and agroecology with activities ranging from literature reviews to FIC courses, scientific experiments, seminars and experiences.

Table 2 - Initial and Continuing Training Courses (FIC), with targets for students, teachers, vulnerable populations, farmers and technicians to be involved in the TRVG and TRRM.

| *Course* | *Target* | *Times/dates* |
|---|---|---|
| Production planning and control assistant in agroecology and organic farming. | -To train family farmers, unions and rural workers, water and forest workers, indigenous people, river dwellers, quilombolas and other traditional peoples. | 160 theoretical Practical, from 05.02.18-16.05.18 |
| Administrator of Community-Based Forestry Enterprises. | -To train 200 ATER technicians, teachers, financial agents, members of rural workers' unions and movements, water and forest leaders, indigenous, riverine, quilombola and gypsy leaders. | 160 practical lectures, from 02.04.18-09.07.18 |

| | | |
|---|---|---|
| Organic horticulturist | -To train 400 family farmers, unions and rural workers, water and forest workers, indigenous people, river dwellers, quilombolas and other traditional peoples. | 160 practical lectures, from 20.08.18-28.11.18 |
| Nurseryman. | -To train 300 family farmers, unions and rural workers, water and forest workers, indigenous people, river dwellers, quilombolas and other traditional peoples. | 160 practical lectures, from 10.05.18-20.08.18 |
| Management of Native Forests for Multiple Use. | -To train 300 family farmers, unions and rural workers, water and forest workers, indigenous people, river dwellers, quilombolas and other traditional peoples. | 160 practical lectures, from 25.06.18-25-09.18 04.02.19-11.05.19 |
| Nutrition attendant. | -Training vocational education students and universities, family farmers, producers in agroecological and organic transition, higher education and federal institute professors, rural youth, ATER agents, in partnership with *slowfood* (http://www. slowfoodbrasil .com/). | 240 practical lectures, from 09.07.18-15.10.18. |

Source: http://pronatec.mec.gov.br/fic/et_gestao_negocios/et_gestao_negocios.php.

Table 3 - Model FIC course plan to be used for training.

| | | | |
|---|---|---|---|
| *Course* | Production planning and control assistant in agroecology and organic farming. | | |
| *Location* | TRVG/TRRM | | |
| *Workload* | 160 or 240 h. | | |
| *Professor* | To be determined. | | |
| *Participants* | Members of rural workers' unions and movements, women of the waters and forests, indigenous leaders, riverside dwellers, quilombolas, gypsy leaders, family farmers, rural teachers, students. | | |
| | ♀ | Indians | No. of young people (up to 29 years old) |
| *A. students:* | 100 | 50 | 50 |
| *Objective* | Carrying out production planning activities in agroecology and organic farming. Interprets the production work plan and programmes line follow-up. Allocates resources. Keeps track of production equipment and the movement of goods. Utilises improvement data. Issues reports. Education: Elementary school incomplete. | | |
| *Methodology* | The courses will be taught in modules I to IV or I to VI, 40 hours each, including 20 hours of *practical activities* in the experiments and at IFRO. There will be lectures and dialogues, technical excursions to the experiments, with the use of audiovisual technologies, handouts and support materials, always with a view to building and valuing professional knowledge. More information: http://cursos.ead.ifro.edu.br/ | | |

Source: http://pronatec.mec.gov.br/fic/et_gestao_negocios/et_gestao_negocios.php.

## 5. Dissemination of scientific work, innovations and audiovisuals

Technical-scientific publications (articles, abstracts for scientific events, book chapters) or didactic-pedagogical publications (booklets, videos, handouts, field days), which will be produced for the priority audience, taking into account the methodology and particularities of the project (Table 4).

Table 4 - Scientific impact, extension, culture, innovation, technology.

| Category | *In* |
|---|---|

| Field day | 1 |
|---|---|
| Technical visit to extractivists | 1 |
| Seminar | 1 |
| Workshop | 2 |
| Scientific article | 1 |
| Abstract in scientific event | 1 |
| Book chapter | 1 |
| e-book book | 1 |
| Website | 1 |
| Research report | 1 |
| Video | 2 |
| Experience | 2 |
| Booklet | 1 |
| Workbook | 1 |
| **Total** → | **21** |

Source: http://pronatec.mec.gov.br/fic/et_gestao_negocios/et_gestao_negocios.php.

## 6. Information on the bidder's technical and managerial capacity

The applicant has a PhD in Forestry Engineering. He is Coordinator of the Project entitled Extension and research centre for rural development

sustainable Rural Territory of Vale do Guaporé-RO, by CNPq/MDA Call No. 11/2014. He works in the areas of Family Farming; Agro-biodiversity and Agro-extractivism; Project Evaluation in Forestry Sciences; Organic Poultry Farming; Animal Climatology; Territorial Development; Solidarity Economy; Rural Extension; Environmental Management; Geographic Web Information; Environmental Legislation; Participatory Methodologies; Poultry and Fish Nutrition; Food Security; Territories of Citizenship. He has published more than 30 scientific articles, more than four expanded abstracts, book chapters and books in the areas mentioned.

## References


BRAZIL. **Law No. 11.326, of 24 July 2006.** Establishes the guidelines for formulating the National Policy for Family Farming and Rural Family Enterprises. Available at:< http://www.planalto.gov.br/ccivil_03/_ato2004- 2006/2006/lei/l11326.htm>. Accessed on: 31 July 2018.

BRAZIL. **Decree No. 7.794, of 20 August 2012.** Establishes the National Policy for Agroecology and Organic Production. Available at: <http://www.planalto.gov.br>. Accessed on: 31 July 2018.

BRAZIL, Federal Constitution. Federal Supreme Court. **Extraordinary appeal**, v. 654432, 2017.

BOUER, R., & Carvalho, M. M. D. (2005). Unique project management methodology: a sufficient condition for project management maturity. **Revista Produção,** 15(3), 347-361.

BORGES, Aurélio Ferreira et al. Agricultural frontier as a factor in the environmental performance of fish farming in Rondônia. **Brazilian Journal of Environmental Education (RevBEA)**, v. 9, n. 1, p. 43-55, 2014.

GOLL, Caroline Kruger Coral et al. Paul Feyerabend's Methodological Anarchism: approaches in Science. **Revista Thema**, v. 15, n. 2, p. 539-552, 2018.

KERZNER, H. **Gestão de projetos**: as melhores práticas. 2.ed. São Paulo: Bookman, 2006.

LACEY, H. The precautionary principle and the autonomy of science. **Scientiae Studia**, v.4, n.3, p.376-392, 2006.

MAYR, E. **Isto é biologia:** a ciência do mundo vivo. São Paulo: Companhia das Letras. 2008. 428p.

PRATES JÚNIOR, Paulo; CUSTÓDIO, Aldo Max; GOMES, Thiago Oliveira. Agroecology: theoretical and epistemological foundations. **Brazilian Journal of Agroecology**, v. 11, n. 3, 2016.

SILVA, M. Evaluation by sar images of the legal reserve of settlements in the state of Rondônia applying the Brazilian forestry code. 2008. 108f. **Dissertation** (Master's Degree in Civil Engineering, Postgraduate Course in Civil Engineering,

Federal University of Santa Catarina.

VARGAS, Ricardo Viana. **Project management**: establishing competitive differentials. 8.ed. São Paulo: Brasport, 2016.

VAZ, DIANA SOUZA SANTOS; BENNEMANN, ROSE MARI. Eating behaviour and eating habits: a review. **Revista Uningá Review**, v. 20, n. 1, 2018.

# CHAPTER 3

## Effective availability of infrastructure and technical support for the development of the project's research activities

Aurelio Ferreira Borges

Edmar da Costa Alves

Mauri Carlos Mazutti

Maria dos Anjos Cunha Silva Borges

Edilberto Fernandes Syryczyk

Diego Soares Carvalho

## Summary

With the slogan Real Food in the Countryside and in the City: for rights and food sovereignty, the project aims to highlight the socio-cultural dimensions of food and nutritional security in order to bring food production and consumption closer together; to build bridges between urban and rural areas; to value agro-biodiversity and organic production, fresh and regional foods, respect for black and indigenous ancestry, Africanity and the traditions of all traditional peoples and communities, and to rescue the identities, memories and food cultures of the population of Rondônia's Territories of Citizenship. It will use the theoretical and practical assumptions of TTs (Technology Transfers), a set of steps that describe the formal transfer of inventions resulting from scientific research by universities and research institutes to the productive sector; Web Map Service, a geographical format specification that aims to provide maps on websites and electronic applications. The original products will be free, as the intention is not to patent the results obtained.

IFRO (Federal Institute of Education, Science and Technology of Rondônia) is part of the programme to expand the Federal Network of Professional, Scientific and Technological Education. IFRO has a virtual learning environment platform, which offers various FIC courses. Information: http://cursos.ead.ifro.edu.br/

The National Programme for Access to Technical Education and Employment (Pronatec) was created by the Federal Government in 2011 with the aim of expanding the supply of professional and technological education courses. Resolution No. 25/Consup/IFRO of 10 July 2015 sets out the Regulations for the Administrative Organisation of the Training Grant actions of the National Programme for Access to Technical Education and Employment

(Pronatec) within the scope of IFRO.

This Resolution states in its Chapter I Art. 2 that the execution of the actions of the Pronatec Training Grant is carried out at IFRO through technical courses and initial and continuing training (FIC), in the face-to-face or distance modalities. § 1° The offer of technical courses and initial and continuing training (FIC), through the campuses, with the aid of the Training Grant must be in accordance with the public indicated by the applicants and articulated with local demands. More information can be found at http://www.ifro.edu.br/site/?page_id=16538.

The Engineering Departments of the National and Foreign Universities involved have laboratories equipped with the software needed to carry out the research. However, the National Universities involved lack a high-performance workstation, compatible with the level of detail and the gigantic volume of information to be processed, which is essential for advancing the line of research in Geosciences for Human Health. The institutions also have computer technicians who will provide effective support for hardware and software issues. In addition, these institutions have professors working in a wide range of fields, such as ecology, geostatistics, cartography, hydrology and soils, who will be able to provide technical and scientific support to the students and researchers involved in the research in question.

UNIR's Geosciences laboratory has granite benches with high stools and can accommodate 40 (forty) students. The main pieces of equipment that make up the laboratory's infrastructure include: 1 crusher for sample preparation, 3 Garmin GPS; King.tools; 8 Brunton compasses; stereo trinocular microscope; 3 microcomputers; 1 ball mill; analytical balance; electromagnetic sieve shaker; hot plate; moisture analyser; stereoscopic microscope; ultrasonic bath; semi-analytical balance; Casa Grande apparatus; Mufla oven; magnetic stirrer with heating; drying oven; digital balance; crusher; jar mill; multimedia projector; distiller; binocular stereomicroscope, 1 camera. Among the consumables are: glassware in general, thermometers, sets of sieves for granulometry, porcelain crucibles, large and small plastic trays, large and small aluminium trays, hammers for sedimentary rock and igneous rock, carving tools of various sizes, 1, 2, 3 and 4 inches, wooden tweezers. The laboratory is used for practical classes, research activities and experiments. The entire laboratory is connected to the internet via points and wireless. This laboratory has a total value of approximately R$ 2,000,000.00.

The Topography and Geography Laboratories at UNIR and IFRO Universities are structured to meet the demands of the geosciences area, offering solid support to Civil Engineering, Medicine and Agronomy courses. Equipped with state-of-the-art instruments that are appropriate to the demand, they allow students and researchers to apply the theoretical concepts learnt in class through practical applications that take place on the university campuses, as well as supporting research and extension activities.

The laboratories have the necessary equipment to carry out the main practical activities related to the areas of topography, geodesy, cartography, photointerpretation, geoprocessing and roads, including: direct distance measurement and indirect distance measurement using laser tape measures and total stations.

The main materials available in the laboratory are:

- Total Stations.
- Dual Frequency GPS receivers (Geodetic).
- optical levels.
- Laser levels.
- Laser tape measure.
- GPS navigation receivers.
- Mirror stereoscope.
- Pocket stereoscopes.
- compasses.
- fibreglass tape measures.
- other surveying accessories, tripods, prism poles, measuring rods, pickets and sledgehammers.
- Topographical maps in analogue format at scales 1:25,000 and 1:50,000; Aerial photographs.

These laboratories have a total value of approximately R$ 4,000,000.00.

## 1. Degree of commitment of companies to the scope of the proposal

- The Energia Sustentável do Brasil consortium of companies, made up of Suez Energy (60 per cent), Eletrosul (20 per cent) and Chesf (20 per cent), manages the Jirau hydroelectric plant, located on the Guaporé/Madeira river.
- The Santo Antônio Energia Consortium of companies, led by Furnas, manages the Santo Antônio Hydroelectric Power Plant, located on the Guaporé/Madeira river.
- The private company Odeon Informática, based in Cuiabá-MT, is interested in developing the interactive e-book Economic and Solidarity Enterprises and the Sigma territory.

These two Consortia and the company are interested and committed to establishing partnerships with Research Centres in order to identify indicators for the scientific, technological, economic, cultural and social impacts that may have arisen following the implementation of this project.

The projects, number of technical and higher education scholarship holders, collaborators and funds raised are listed below (Table 1).

Table 1 - IFRO internal calls for proposals that have supported research projects since 2016, with scholarship holders and funds raised.

| *Public notices* | *Research project* | *Technical Fellow* | *Senior Fellow* | *Scholarship holder* | *Collaboration Technical education* | *Higher education collaboration* | *Funds raised* |
|---|---|---|---|---|---|---|---|
| 13 | 75 | 43 | 55 | 7 | 2 | 18 | R$ 150.000,00 |

Source: https://drive.google.eom/a/ifro.edu.br/file.

## 2. The teaching unit's institutional curriculum

The IFRO Cacoal Campus has partnerships with public and private teaching, research, extension and production institutions that co-operate with the Campus. To date, the aim has been to organise exchanges and supervised internships for technical and higher education students. The partner public institutions are: Embrapa (Brazilian Agricultural Research Corporation), Emater (State Company for Rural Assistance and Extension), Idaron (Agrosilvopastoral Health Defence Agency), Sedam (State Secretariat for Environmental Development), Ceplac (Commission for the Execution of the Cocoa Farming Plan), INPA (Institute for the Protection of the Environment).

National Amazon Research Centre), SEAGRI (Secretariat of Agriculture, Livestock and Land Regularisation), SEBRAE (Support Service for Micro and Small Businesses in Rondônia), State Secretariat for Environmental Development (SEDAM), SIPAM (Amazon Protection System), Colorado do Oeste City Hall, Petrolina-PE City Hall, two Dinter PhD courses with agreements between Unesp and UFF.

It has technical and scientific partnerships with several international universities: National University of Colombia, University of Beni, Portuguese Universities, Technological Institute of Costa Rica University of Paris III.

Private companies include those active in agricultural production, such as farms and agricultural centres, as well as companies that produce and process agricultural products, such as slaughterhouses and dairies. Other partnerships will be sought and implemented as and when they become necessary for the development of the project.

The companies that currently have agreements with the Campus are: Fazenda Fertipar, Agrocampo, Agrocat - Catini, Marquetti & Cia Ltda - Cerejeiras, Agrocat - Kleber José Marim Silva, Agrofértil, Agroindústria e Piscicultura Santa Clara Ltda, Agromaza - Agrop. Martins da Amazônia Ltda, Agropecuária Boa Safra, Agropecuária Comercial e Industrial Caarapó S/A. The following are the technical, undergraduate and postgraduate courses offered by IFRO:

## Integrated Technicians

- Agroecology Technician (Cacoal).
- Agricultural Technician (Ariquemes, Cacoal, Colorado do Oeste).
- Food Technician (Ariquemes).
- Building Technician (Porto Velho Calama, Vilhena).
- Electromechanics Technician (Vilhena).
- Electrical Technician (Porto Velho Calama).
- Forestry Technician (Ji-Paraná).
- Computer Technician (Ji-Paraná, Porto Velho Calama, Vilhena).
- Computer Maintenance and Support Technician (Ariquemes).
- Chemistry Technician (Ji-Paraná, Porto Velho Calama).

## Subsequent Technicians

- Agricultural Technician (Cacoal).

- Building Technician (Porto Velho Calama, Vilhena).
- Electromechanics Technician (Vilhena).
- Electrical Technician (Porto Velho Calama).
- Finance Technician (Porto Velho North Zone).
- Computer Technician (Ji-Paraná).
- Internet Computing Technician (Porto Velho North Zone).
- Computer Maintenance and Support Technician (Porto Velho Calama, Vilhena).

*-Concomitant Technicians*

- Finance Technician (Porto Velho North Zone).
- Internet Computing Technician (Porto Velho North Zone).
- Computer Maintenance and Support Technician (Guajará-Mirim).

## Graduation

*-Degree*

*L* Degree in Biological Sciences (Ariquemes, Colorado do Oeste).

- Maths degree (Cacoal, Vilhena).
- Degree in Chemistry (Ji-Paraná).
- Degree in Physics (Porto Velho Calama).

*-Technologist*

- Higher Technology Course (CST) in Environmental Management (Colorado do Oeste)
- CST in Dairy Products (Colorado do Oeste)
- CST in Systems Analysis and Development (Ji-Paraná, Porto Velho Calama and Vilhena)
- CST in Public Management (Porto Velho north zone)

*-Engineering*

- Agricultural Engineering (Colorado do Oeste).

*-Bachelor*

- Zootechnics (Colorado do Oeste and Cacoal).

Table 2 - Lato sensu postgraduate courses offered by IFRO in areas related to rural, regional and territorial development.

| *Name of lato sensu postgraduate course* | *Workload* |
|---|---|
| Environmental management. | 360 |
| Youth and Adult Education. | 360 |
| Georeferencing and Geoprocessing. | 360 |
| Informatics in Education. | 360 |

Source: www.ifro.edu.br.

## 3. Characterisation and role of partner institutions in the project

- **Facimed** (Faculdade de Ciências Biomédicas de Cacoal-RO): will participate as a technical-scientific collaboration through research professors and human medicine students - partnerships already signed.

- **Una** (National University of Colombia): will participate as a technical-scientific collaboration through research professors and students from the doctorate in Agroecology and Production - partnerships already signed.

- **MAPA** (Ministry of Agriculture, Livestock and Food Supply): through the Organic Production Commission, it will provide guidance on the duties defined in Normative Instruction No. 13 of 28 May 2015, such as coordinating actions and projects to promote organic production; suggesting adjustments to production standards and organic quality control; assisting in inspection through social control; and proposing public policies for the development of organic production in the TRVG and TRRM.

The project "Agroecological Factsheets: Appropriate Technologies for Organic Production" will be part of this project, which aims to provide technical information on appropriate technologies for organic production systems, in a summarised form, in simple and accessible language for traditional peoples and communities and rural producers. Information: http://www.agricultura.gov.br.

- FUNAI (National Indian Foundation): discussions between this project and FUNAI will focus on creating and improving legal mechanisms to prevent indigenous populations in the TRVG and TRRM from being expropriated of their rich intellectual heritage, produced over generations. The obvious problems of knowledge associated with biodiversity have been the target of cases of biopiracy. International agreements such as the Convention on Biological Diversity and Agenda 21 have highlighted the urgency of the problem.

- **MDA** (Ministry of Agrarian Development): will work by adjusting procedures and building, in a participatory manner, the routine for evaluating, processing, contracting and monitoring Ater projects currently developed by the project coordinated by the proponent, in accordance with process 463352/2014-9, CNPq/MDA Call No. 11/2014.

-Embrapa**:** will jointly manage the "Participatory guarantee system for organic products" with this project. Brazilian legislation provides for three different ways of guaranteeing the organic quality of its products: Certification, Participatory Guarantee Systems and Social Control for Direct Sales without Certification. The so-called Participatory Guarantee Systems, together with Certification, make up the Brazilian Organic Conformity Assessment System - SisOrg. For their proper functioning, Participatory Guarantee Systems are characterised by Social Control and Solidarity Responsibility, which makes it possible to generate the credibility appropriate to different social, cultural, political, institutional, organisational and economic realities.

Information:///D:/Users/AURELIO/Downloads/sistema_participativo.pdf.

- **INCRA** (National Institute for Colonisation and Agrarian Reform): with the support of this proposal, this institute will implement the basic infrastructure needed in the TRVG and TRRM agrarian reform areas, both directly and in partnership with other government bodies. The priorities will be the construction and completion of side roads and basic sanitation, through the implementation of water supply and sewage systems, as well as the construction of rural electrification networks, with the aim of providing the physical conditions necessary for the sustainable development of the settlements and their traditional communities.

The project will collaborate in the revision of the Normative Instruction that regulates the procedure for the identification, recognition, delimitation, demarcation, titling and registration of lands occupied by remnants of quilombola communities in the TRVG and TRRM, as it is understood that this hinders the process of demarcating quilombola lands; and

(2) co-participation in the First National Strategic Titling Plan aimed at finalising the administrative procedures for titling quilombola territories initiated by INCRA in the TRVG and TRRM, with an indication of the short, medium and long-term actions required to remedy the demand.

More information at http://www.incra.gov.br/infraestrutura_assentamentos.

-Slow **food**: the basic principle of this movement is the right to enjoy food, using artisanal products of special quality, produced in a way that respects both the environment and the people responsible for production, the producers.

Through its gastronomic expertise related to politics, agriculture and the environment, Slow Food has become an active voice in agriculture and ecology. Slow Food combines pleasure and food with awareness and responsibility. The association's activities with this project will aim to defend biodiversity in the food distribution chain, spread taste education, and bring producers and consumers of speciality foods closer together through events and initiatives. More information at http://www.slowfoodbrasil.com/slowfood/o- movimento.

**-SEPPIR** (Special Secretariat for Policies to Promote Racial Equality), of the Ministry of Justice and Citizenship. Federal policy for quilombos is linked to the Brazil Quilombola Programme (PBQ), coordinated by SEPPIR. This programme was launched in 2004 with the aim of consolidating the frameworks of state policy for quilombola areas. As a result, the Quilombola Social Agenda (Decree 6261/2007) was instituted, which groups the actions of various ministries aimed at the communities into four main axes, namely: 1) Access to Land; 2) Infrastructure and Quality of Life; 3) Productive Inclusion and Local Development; and 4) Rights and Citizenship.

SEPPIR will support the consolidation of state policy frameworks for Quilombos. More information at http://www.incra.gov.br/quilombola

**-Icmbio/Mma** (Chico Mendes Institute for Biodiversity Conservation of the Ministry of the Environment): the National Programme for Access to Technical Education (Pronatec) Bolsa Verde FIC courses, run by IFRO in partnership with the MMA and this project will train residents of areas included in the Bolsa Verde Programme in activities that will contribute to raising their income and the sustainable use of natural resources. Bolsa Verde will help people living in extreme poverty who live in conservation units and environmentally differentiated settlements in the TRVG and TRRM, with an incentive of R$300 every three months.

The partnership will also involve geoprocessing for the areas of traditional peoples and communities in these two Territories. This section will present the geoprocessing tools, services and functionalities developed by ICMBio to present spatial data and information relevant to the scope of this project. The products presented will be developed with reference to the Federal Government standards established by Decree No. 6.666 of 27 November 2008, which establishes the National Spatial Data Infrastructure - NSDI and the Electronic Government Interoperability Standards - ePING.

More information at http://www.icmbio.gov.br/portal/

**-CONAB** (National Supply Company): using information from this project, the company will carry out studies and statistics on the prices of agroecological and organic products in these two territories, as well as surveying production costs, planting and harvest expectations, and the volume and location of public and private stocks of a range of products.

With regard to defining public policies for food supply, within the scope of the Minimum Price Guarantee Policy (PGPM), the Company will be responsible for its implementation. This will be done through Federal Government Acquisition (AGF), an instrument capable of balancing the income of rural producers from traditional peoples and communities, family farmers and their cooperatives, in the face of fluctuating market prices.

More information can be found at http://www.conab.gov.br/index.php

**-Unir** (Federal University of Rondônia): will take part in technical and scientific collaboration through research professors and postgraduate students in Sustainable Rural Development - partnerships already signed.

**-Ufrrj**: will participate as a technical-scientific collaboration through research professors and postgraduate students in Sustainable Rural Development - partnerships already signed.

## 4. Didactic-pedagogical orientation and methodologies applied

For the execution of the technological extension, professional education and research activities to be carried out.

According to Article 3 of Decree 5.154/2004 (BRASIL, 2004), which regulates Chapter III of the LDB, "Initial and Continuing Training courses and programmes for workers, including training, improvement, specialisation and updating, in all schooling conditions, may be offered according to training itineraries, aiming to develop skills for productive and social life".

The Initial and Continuing Training courses and programmes in this proposal aim to:

*P* To provide workers with the development of skills for productive and social life.

. To promote the training, improvement, specialisation and updating of professionals in the areas of professional and technological education.

. To qualify and re-qualify workers, preparing them to engage in a type of professional activity in order to promote their entry and/or re-entry into the labour market.

. Expanding the professional competences of workers.

. To awaken citizens' interest in re-entering school, in courses and programmes that promote higher education and increased socio-environmental awareness.

## 5. Methodology and teaching materials for FIC courses

Methodology

The Initial and Continuing Training (FIC) courses will be taught in classes that articulate practical and academic knowledge in a complementary relationship, in which the students' process of appropriating knowledge allows for theoretical and practical improvement. These will be lectures and dialogues, using audiovisual technology, handouts and support materials, always with a view to building knowledge by valuing professional knowledge. It should be emphasised that the theoretical contributions worked on in class must make sense in the reality in question.

The courses will provide conditions for students to develop professional competences through theoretical studies, case discussions, debates, games and experiences, simulations of professional practices, problem-solving, reflection on videos, participation in lectures, among other activities that require the active involvement of civil servants/students and stimulate criticism, creativity and decision-making.

Teaching and learning materials

For each module, there will be a workbook prepared by the teachers with guidance and supervision from the course coordinator. The material should summarise the theories and provide references for in-depth bibliographic searches, as well as examples, case studies and exercises that encourage reflection on professional practice. The handouts will be distributed in physical, printed or digital form.

Learning assessment

Student assessment will be carried out as an integral part of the educational process and will take place throughout the course so as to allow reflection-action-reflection on learning and the appropriation of knowledge, emphasising its diagnostic, formative, procedural and summative dimensions. During the educational process, the teacher should be attentive to the student's effective participation by observing attendance, punctuality and involvement in work and discussions.

Means of operationalising the evaluation will be considered:

- seminars.
- individual and group work.
- written and oral tests.
- demonstration of techniques and laboratory.
- dramatisation.
- presentation of work.
- portfolios.
- reviews.
- self-evaluation, among others.

Students who attend the course more than 75% of the time and achieve 60% of the proposed activities will be considered fit. For registration purposes, the concept of Apt (A) will be used for civil servants/students who meet the criteria established and mentioned above, and Not Apt (NA) for civil servants/students who do not.

This Improvement proposal aims to modernise the TRVG and TRRM NEDETs and is designed to expand and consolidate a reference network in technological extension, linked to professional and technological education aimed at social and productive inclusion.

Through extension methodologies[7] , linked to applied research and vocational and technological education, the aim is to expand opportunities for productive inclusion; to strengthen local production systems through the development of technical and technological solutions; income generation and occupational integration through technological extension methodologies geared towards local and sustainable development.

In order to fulfil the theoretical and practical scope of this proposal, this teaching, extension and research centre will be set up as a technological extension unit for the development, promotion and supply of services and products aimed at strengthening local production systems. The specific aims of this proposal are:

a) Promote inclusion and sustainable social development through the articulation of teaching,

---

[7] Extension is an interdisciplinary, educational, cultural, scientific and political process that promotes transformative interaction between the university and other sectors of society, guided by the constitutional principle of inseparability from teaching and research. It has an organic-institutional character, integration into the territory and/or population groups, clarity of guidelines and orientation towards a common objective, being carried out in the medium and long term (DE FREITAS, 2016).

research and extension actions, especially the promotion of technological extension articulated with applied research and professional and technological education.

b) Promote the articulation between applied research, technological extension and technological and professional education aimed at developing local economic vocations and improving the quality of life, especially of people living in extreme poverty and on low incomes.

c) Improve and develop knowledge, techniques and social technologies based on local demands and participatory methodologies, with a focus on improving local and regional production systems.

d) Support actions to democratise information and popularise science and technology. e) Promote the coordination of the various local and regional social actors through the proposal for shared management of the Centre.

e) It is also important to mention the interface with the National Plan for Rural and Sustainable Development (PNDRSS), which has as one of its objectives the consolidation and strengthening of international, regional and multilateral spaces, the rural development agenda with an emphasis on family farming and agroecology, as well as the promotion of ethno-development, valuing agro-biodiversity and socio-biodiversity products, in addition to its specific objectives aimed at valuing young people and women.

## 6. Project research activities

A - Create the Mother Bank (UBS) of Food, Oilseed, Fibre, Fodder and Green Fertiliser Seeds at the Rural Family Production Unit (URPF), the São Miguelense Rural Association for Mutual Aid, which would receive an annual quota of seeds from the experiments carried out in the backyards of the associations involved and in the experiments of the IFRO Campuses in the municipalities of Cacoal, Colorado do Oeste, Ariquemes and Porto Velho. This bank would be used to store the seeds, which would come from each Community Bank of these communities of traditional peoples in the TRVG and TRRM.

In order to carry out the research, the case study method will be used, given the possibilities of construction and reconstruction, focusing on the spatial and temporal dimensions that are clearly delimited and linked to the object of study. this method consists of an intensive examination of the unit of study, using all available techniques. The focus of this study will be traditional knowledge, the use and management of seed banks by traditional peoples and communities. The approach used will be a systemic one, as pointed out by MORIN (2014), which seeks to understand not only the whole-parts relationship, but also to understand the environment as a macro-organisation, an organisation that is both ecological and social and in a continuous process of recurring reorganisation through interactions. The concept of system, for the author, consists of three parts: (1) system - which expresses the complex unity and phenomenal character of the whole, as well as the complex of relationships between the whole and the parts; (2) interaction - which expresses the set of relationships, actions and feedbacks that take place and are woven into a system; (3) organisation - which expresses the constitutive character of these interactions - that which forms, maintains, protects, regulates, governs and regenerates and which gives the idea of the system its phenomenological essence.

The sample will be made up of vulnerable populations in the TRVG. The sample size will be with a confidence level of 97 per cent and an error of 3 per cent.

The suitability of the sample for factor analysis will be assessed using the Bartlett's sphericity and Kaiser-Meyer-Olkin (KMO) tests.

Quantitative and qualitative approaches will be used in the scientific research. Data will be collected from vulnerable populations using questionnaires and evaluation of the data collected. Evaluations will be carried out using Sigma Territory and SPSS software. Parametric or non-parametric tests may be used to evaluate the data collected (Table 3).

Table 3 - Statistical perspectives for evaluating the data.

| N° of Samples | Dependency | Parametric tests | *Non-Parametric Tests* |
|---|---|---|---|
| 2 | - Yes | - paired Student's t | - Wilcoxon |
| | - No | - unpaired Student's t | - Mann-Whitney |
| > 2 | - Yes | - ANOVA repeated measures | - Friedman |
| | - No | -ANOVA | - Kruskall-Wallis |

Source: the authors.

The techniques used in this study will be: a questionnaire, open-ended interviews, informal conversations, a field diary and participant observation.

The results obtained will be published and disseminated in workshops, booklets, seminars, lectures, videos, field days, books, articles, experiences, etc. B - Monitor a prototype experimental unit of the agroforestry model adopted for silviagriculture, defined by MACEDO (2000) as the combination of trees, shrubs or palms with agricultural species. The experiment will take place in the Calcário Extractive Reserve in Costa Marques. The development of native forest species and introduced species will be assessed, comparing them to the conventional planting model. The statistical model adopted will be a factorial 2 (treatments: conventional planting of forest species and planting in an agroforestry system) x 6 (species: açaí, cassava, pupunha, mahogany, copaiba, pineapple) with 3 replications, following a completely randomised design (Figure 1).

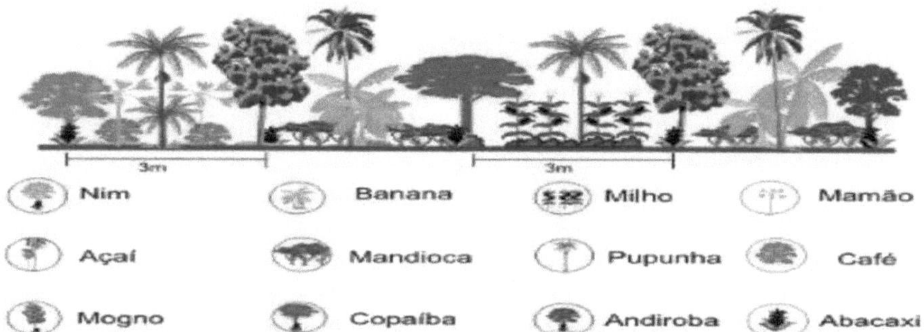

Figure 1 - Model of the experimental silvicultural agroforestry unit.

Source: www.ekoe.com.br.

In this experiment, the species pupunha, mahogany, copaiba, andiroba will be planted in 3x2m rows; in between the rows, neem, banana, corn, papaya, açaí, coffee and pineapple; in between the rows, during the month of September 2017, in an area of 0.012 ha, 1 kg of Phaseolus vulgaris L. (pigeonpea) of the canary variety and 10 units of Manihot utilissima Pohl (cassava) will be planted. The statistical model adopted will be a factorial 2 (treatments: conventional planting of forest species and planting in an agroforestry system) x 3 (native tree species), with 3 replications per species, analysed using the T-test ($P > 0.05$), following a completely randomised design (DIC). The averages for height, diameter at ground level and crown area of the tree species will be compared using the Scott-Knott test.

The results obtained will be published in workshops, booklets, seminars, lectures, videos, field days, books, articles, experiences, etc. C - Determine the *Nurmalizo p Differenco Vugetatido Index,* used to determine land cover in a GIS (Geographic Information System). For this purpose, satellite images from the TM Landsat-5, ETM+ Landsat-7 CCD CBERS-2 and CCD and HRC CBERS 2B sensors are used. Remote sensing and geographic information system data integrated with soil loss prediction models are important tools for assessing the environment, such as mapping and quantifying areas of agroecological production of fruit, vegetables and legumes on indigenous lands in family farming. Also the issue of water availability, by mapping springs, estimating land cover, estimating biomass and forecasting crops from agroecological and organic production.

The scale adopted in this work will be 1:50,000, defined on the basis of the planialtimetric and hydrology maps generated by the IBGE (Brazilian Institute of Geography and Statistics) for the study region, which was selected due to the spatial and temporal diversity of its coverage, associated with the diversity of soil classes and slopes.

All the images will be obtained free of charge from INPE (National Institute for Special Research) via the website www.dgi.inpe.br/cdsr/ and with the following characteristics: Landsat-5 TM sensor, orbit 217, point 076; Landsat-7 ETM+ sensor, orbit 217 and point 076; CBERS 2 CCD sensor, orbit 151 and point 125; and CBERS 2B CCD sensor, orbit 151 and point 125, showing the list of images and the respective dates of acquisition, according to the theory proposed by Durigon (2011).

The image bands will be transformed into 8-bit raw format (without headers) using the Spring 4.3.2 software, developed by INPE, and then processed by the 6S model, whose output consisted of images with surface reflectance values (in 8-bit format), ready to be georeferenced.

All the bands of the images used will be treated, with the exception of the Cbers-2B images, as there is no suitable methodology for correcting atmospheric effects due to the sensor's lack of reliable calibration. The parameters used to correct the bands will be the longitude and latitude of the centre of the micro-basin under study, the average altitude of the study area, the tropical atmosphere model, the continental aerosol model and horizontal visibility of 16 km.

The statistical programme that will be developed in the experiment will be free. It is not intended to be patented.

D - This experiment at the Cacoal Campus (IFRO) aims to present a product (Figure 6) that uses the concept of environmental sustainability to help combat environmental problems affecting the TRVG and TRRM territories. The product uses recyclable materials to produce a solar panel, called an ASAR (Alternative Solar Reflection Heater), which heats water for use in the rural homes of traditional peoples and communities.

The CAD/CAE systems methodology will be used. In this sense, the article proposes to design a low-cost solar heater that meets the needs of traditional peoples and communities in these two Territories. NAKAMURA et. al. 2003, cites CAD (Computed Aided Design) and CAE (Computer Aided Engineering) systems as important tools for making product design feasible in short timeframes, offering the opportunity for simulation and reducing costs during the product development phase.

CAD is software for technical drawing, bringing together various tools for a wide variety of purposes.

The CAE system is a tool that analyses and processes calculations in such a way as to minimise the engineer's manual efforts, worrying less about the operational side and more about the strategic, making CAE a powerful tool for reducing project costs and minimising the time it takes to launch a product.

The CAD platform used *Solidworks* for three-dimensional modelling of the prototype and the CAE software used *Comsol* for heat exchange simulations.

Examples of collectors are those made from recyclable materials such as PET bottles, milk cartons, PVC ceiling tiles and even fluorescent tubes. In the case of this project, the difference is the use of aluminium cans (Figure 2).

**Figure 2 - Solar collector made from PVC pipes.**
**Source: http://www.sempresustentavel.com.br/solar/aquecedor/aquecedor-solar.htm.**

This is a can collector with approximate dimensions of Im x lm, made up of an arrangement of 7 x 11 cans in a total of 77. The system also has a piece of wood painted black in order to absorb heat. A transparent plate closes the system to stimulate the greenhouse effect and thus favour an increase in water temperature. In the panel, the sun's rays are reflected (the phenomenon of reflection) by the aluminium cans, which are directed to the PVC pipes at the focus of the cans to heat the water, which receives the energy in the form of heat. The heated water is stored in a boiler tank. The user can regulate the temperature by means of a cold water valve (Figure 3).

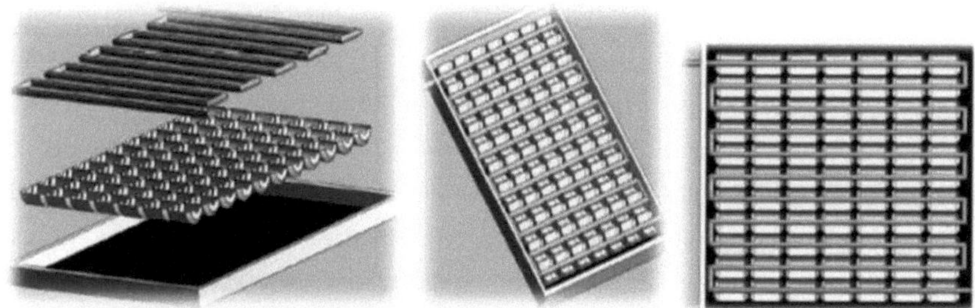

**Figure 3 - Modelling a solar panel using Solidworks.**

**Source: http://revistageintec.net/portal/index.php/revista/article/view/297/347.**

The results obtained will be published and disseminated in workshops, booklets, seminars, lectures, videos, field days, books, articles, experiences, etc. E - Experiment with a photovoltaic panel. This is the element responsible for converting solar energy into electrical energy. The model S 55 solar panel with 36 monocrystalline silicon cells, matrix 4 x 9 will be used in the experiment to be installed in the Caltário River Extractive Reserve, located in the municipality of Costa Marques-RO (Figure 8). The photovoltaic panel generates energy from small silicon cells with the sun as the generating source. The energy is captured on its surface when the sun's rays hit it at different angles, the intensity of the beam being greater when the incidence is at 90° (DANTAS, 2012).

Figure 4 - Model S 55 solar panel.

Source: RIBEIRO (2010).

From the author's perspective, the advantages of photovoltaic (PV) panels compared to other forms of electricity generation are manifold. The advantages include: a) an unlimited resource of solar energy; b) availability in all parts of the world; c) modularity, ranging from milliwatts in consumer products to gigawatts in future power stations; d) during operation, it produces electricity without emitting pollutants or producing atmospheric waste; e) quietness during operation; f) a technically proven service life of more than 30 years; g) low maintenance costs; h) no fuel costs; i) decentralisation of the electricity grid.

The methodology used will be to evaluate the management model developed for communities of traditional peoples who will use two photovoltaic energy module models; the results of the implementation of quality of service standards; the monitoring of user satisfaction levels and failure and service interruption rates; and the creation of subsidies for electrification programmes for isolated communities.

The project's partners will be the Mamirauá Sustainable Development Institute (IDSM) and the Winrock Brazil International Institute. The results obtained will be published and publicised in workshops, booklets, seminars, lectures, videos, field days, books, articles, experiences, etc.

F - Experiment to collect rainwater in mini cisterns, to be installed in the Uru-Eu-WAu-WAu, Cinta Larga and Paiteré Cará Indigenous Land Association Communities. The system takes into account the water that comes from the conductors, passes through the filters, goes to the disposal reservoir and then on to the storage reservoir (Figure 5).

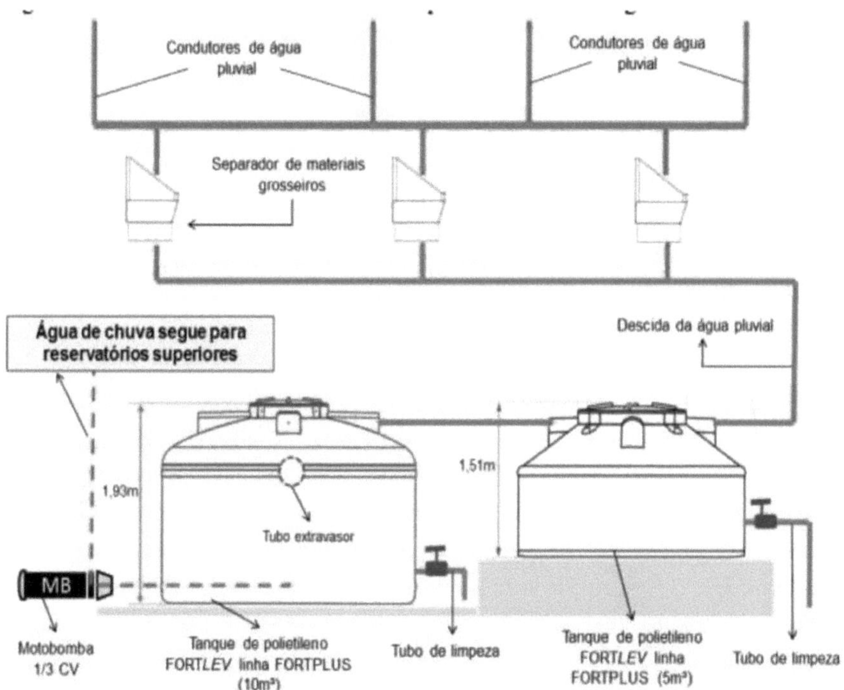


Figure 5 - Mini rainwater harvesting cistern methodology.
Source: https://repositorio.ufsc.br.

Methodology: The control and monitoring of water for reuse will be carried out taking into account the physical-chemical and biological parameters recommended by CONAMA Resolution 357/05 and Ministry of Health Ordinance 518/2004. The following parameters will be assessed (NMP/100 ml), Colif. Total (MPN/100 ml), number of Aedes aegypti larvae, among others (Table 4).

Table 4 - Parameters for evaluating water from the mini cistern.

| *Parameters* | *Sample I* | *Sample II* |
|---|---|---|
| Total Coliform (MPN/100 ml) | | |
| Oils and greases (mg/l) | | |
| pH | | |
| COD (mg/l) | | |
| Suspended Solids (mg/l) | | |
| Sedimentable solids (ml/l) | | |

Source: the authors.

The results obtained will be published and disseminated in workshops, booklets, seminars, lectures, videos, field days, books, articles, experiences, among others, to be held in the TRVG and TRRM territories.

G - Implement the "Entrepreneurship, Territory and Innovation" experiment at the Quilombola Remnant Association of Pedras Negras do Guaporé, in the municipality of São Francisco do Guaporé and at IFRO, Campus Cacoal.

In the floodplain areas of the TRVG and TRRM, during the river's ebb (August, September and October) and dry (November, December and January) periods, vegetables are grown in the open on beds made directly in the ground, with some species being grown several times during these periods. During the periods of flooding (February, March and April) and inundation (May, June and July), some producers continue to grow vegetables in suspended beds in the high floodplains and in areas that are not flooded.

NODA et al. (2007) considered growing crops in suspended beds as a strategy for diversifying and expanding the productive capacity of the production system of floodplain family farmers in the Amazon. In these structures, the plants are free from excess humidity during the rainy season and cultivation does not need to be interrupted during the flood period. The hanging beds are made of wood, like a box, crate or tray, 1.0-1.2 metres wide, 20 cm high (deep) and of varying lengths, held above the ground depending on the amount of water at the peak of the flood. They should have holes at the bottom to allow rainwater and irrigation water to drain away, and be filled with floodplain soil and animal manure or decomposed wood, known as paú.

Olericulture technicians have pointed out that improving this system requires adapting both the structure and the substrates for the growing bed. The studies should culminate in projects to define different materials (resistant and economically viable) that can replace the use of wood in the construction of structures, given the environmental restrictions. The substitute materials should not generate waste that pollutes the environment. The substrates should be obtained from regional raw materials, be suitable for vegetable production and not cause damage in terms of environmental pollution (KANO et al., 2013).

*Growbed* Grande is a solution for growing vegetables for traditional peoples and communities. The product is part of a new vision of the food production cycle and aims to make it easier to grow food in rural and urban environments. Dimensions: height: 35 cm; width: 125 cm; length: 125 cm; Weight: 14 kg (Figure 6).

Figure 6 - Technique for growing vegetables in recyclable plastic bags.

Source: www.ekoe.com.br.

The main benefits of the product are:- high productivity (up to 50 per cent higher than an ordinary vegetable garden):- easy to maintain;- saves water (80 per cent less water use than an ordinary vegetable garden);- saves time (water tank lasts up to 3 weeks);- healthier plants;- complete and easy-to-assemble vegetable garden.

-The *experiment:* Survival assessment of seedlings of cultivated vegetables lettuce *(Lactuca sativa),* spring onion *(Allium fistulosum),* chicory *(Erygium foetidum), coriander (Coriandrum sativum),* cabbage *(Brassica oleracea* var. *acephala),* maxixe *(Cucumis anguria),* peppers *(Capsicum annuum*), chillies *(Capsicum* spp.), tomatoes *(Solanum lycopersicon)*, grown in a Growbed Grande system and in a vertical panel with hanging containers (Figure 7).

Figure 7 - Vertical panel with vegetables.

Source: www.ekoe.com.br.

a) vertical panel system with hanging containers: the matrices will be grown in containers with a capacity of 5.0 litres filled with a mixture of organic soil and organic substrate (made from pine bark and vermiculite), in a ratio of 1:1. The containers will be kept on supports at a height of 1.50 m, spaced 40 cm apart, in a greenhouse covered with low-density polythene with UV additives;

b) *Growbed* Grande: the nine (9) vegetables will be grown in beds with a height of 35 cm, a width of 125 cm and a length of 125 cm, placed inside the greenhouse at a spacing of 1.0 m between beds. The beds will be previously prepared with a mixture of organic soil and organic substrate in a 5:3 ratio. Ten mother plants of each of the nine vegetables will be used per Growbed Grande.

Drip irrigation will be used in both systems. The other cultural treatments, including the phytosanitary management of the mother plants, will be carried out according to the crop's needs. Once the stoloniferous strands have emerged, in the conventional system, the plants will be kept to develop on the beds, allowing the seedlings to root. In the hanging pot system, the plants will be kept upright and suspended.

The seedlings from these systems will be collected and planted in 128-cell expanded polystyrene trays filled with vermiculite and taken to a mist chamber for 15 days to root. After this period, they will be transplanted to the production beds.

Two analyses of the percentage of seedling survival will be carried out. The first will take place when the seedlings are transplanted into the beds, and the survival of the seedlings will be analysed 15 days after planting in the expanded polystyrene trays. The second analysis will be carried out 15 days after transplanting to the final growing site, i.e. 30 days after the first planting of the seedlings.

To assess dry matter (g kg-1), the plant tissue will be dried in an air-flow oven at 65°C for 72 hours; the total carbohydrate content (mg 100 mL-1) will also be assessed, following the method described by DUBOIS et al. (1956). Samples will be taken only once when the seedlings are collected (hanging pot system and conventional) and these will be sent to the analysis laboratory at the Centre for Nuclear Energy in Agriculture (CENA/USP). Each sample will consist of 5 plants, from which the aerial parts will be removed, leaving only the roots to be analysed.

The dead seedlings from both systems will also be analysed at the Phytopathology Clinic of the Department of Entomology, Phytopathology and Zoology at ESALQ/USP. The analyses will involve observing the samples under an optical microscope and visually identifying the causal agent of the diseases by their vegetative or reproductive structures.

For the mapping experiment, the experimental design will be entirely randomised, in a 9x2 factorial scheme (nine cultivars and two seedling production systems), with three replications of 5 seedlings. The results obtained will be subjected to analysis of variance (F test) and the means will be compared using the Tukey test at a 5% probability level. The results obtained in percentage values will be transformed to arc-sen . However, the averages shown in the table of results will be the original ones.

The results obtained will be published and disseminated in workshops, booklets, seminars, lectures, videos, field days, books, articles, experiences, etc.

## 7. Products expected as a result of the research project

Expected results:

a) Mapping and quantifying areas of agroecological production of fruit, vegetables and legumes on indigenous lands, family farms, areas of traditional peoples and communities (women of the waters and forests, rural women, rural youth, quilombos) living in rural areas and in situations of extreme poverty in the TRVG and TRRM.
b) Number of trained ATER technicians.
c) Number of registrations of traditional peoples and communities organised in groups, to be included in the National Register of Organic Products.
d) Number of traditional peoples in these two Citizenship Territories who have obtained access to genetic resources and socio-biodiversity products.
e) Number of community seed banks and Seed Processing Units of interest to Agroecology and Organic Production created within the scope of the project.
f) Seven FIC courses.
g) Eight lectures.
h) Two professional experiences.
i) Two workshops.
j) A seminar.
k) Eight experiments.

l) A field day.
m) Research results from the experiments.
n) Publication of a scientific article.
o) Website: agroecological products and organic production
p) Publication of a book chapter.
q) Publication of a book, workbook, booklet and video.

Products:

a) Estimation of land cover, biomass estimates and crop forecasts for agroecological and organic production.

b) Process of determining the spectral bands of the Spot Vegetation sensor in the red (0.65pm to 0.72pm) and infrared (0.7pm and 1.3pm).

c) Calculate the Standardised Vegetation Index (SVI), which quantifies NDVI anomalies

d) Field day.

e) Technical visit to extractivists.

f) Seminars.

g) Workshops.

h) Scientific article.

g) Abstracts for scientific events.

h) Book chapter.

i) Patent e-book.

j) Website.

k) Research reports.

l) Video.

m) Experience.

n) Booklet.

o) Workbook.

p) FIC courses (Initial and Continuing Training, minimum 160 hours).

q) Results and Technology Transfer of scientific research into community organic gardens, rainwater collectors, photovoltaic panels for traditional populations, wind energy in traditional communities, cartography for vulnerable populations, solar heaters, community banks of Creole and traditional seeds, agroforestry and organic silviagriculture models and research into determining a Geographic Information System for Agroecology and Production. Organic.

After 24 or 30 months, the project will be institutionalised by raising funds for capital costs, funding and scholarships through institutionalisation funds from IFRO, UNIR and FAPERO. It is intended that this project will be extended to the other five (5) Rural and Citizenship Territories in Rondônia for at least another six (6) years.

## 8. Detailed budget.

Table 5 - Detailed budget for capital, funding and grants.

| **Capital Expenditure** | | | | |
|---|---|---|---|---|
| *Items of equipment and permanent material* | | | | |
| **Itemisation** | Unit | Quantity | Unit R$ | Total value R$ |
| Bac 100 handle/needle steriliser. | Unit | 01 | 1.200,00 | 1.200,00 |
| Infrared laser thermometer gm 200. | Unit | 01 | 170,00 | 170,00 |
| Thermos box without thermometer, 8.5 | Unit | 01 | 210,00 | 210,00 |

| litres | | | | |
|---|---|---|---|---|
| 8x2mm sieve,[3] /4. | Unit | 06 | 250,00 | 1.500,00 |
| Blood collection chair, FMA 157. | Unit | 02 | 370,00 | 740,00 |
| Digital scale 0.01 x 3,200 g. | Unit | 01 | 2.134,00 | 2.134,00 |
| Glass repipette, 0.1-20 ml. | Unit | 01 | 395,00 | 395,00 |
| Bench phantom 0 to 14 phb 500. | Unit | 01 | 1.558,00 | 1.558,00 |
| Paediatric scale, 16 kg 063 la. | Unit | 02 | 695,00 | 1.390,00 |
| Mod 3.200 mechanical watch. | Unit | 01 | 58,00 | 58,00 |
| K30-004 4-channel digital timer. | Unit | 01 | 170,00 | 170,00 |
| H4Q incubation tray. | Unit | 01 | 420,00 | 420,00 |
| Stand for staining 32 slides, stainless steel. | Unit | 02 | 200,00 | 400,00 |
| Data show. | Unit | 01 | 2.000,00 | 2.000,00 |
| XL digital compass chronometer. | Unit | 01 | 39,00 | 39,00 |
| Portable digital scale 1 g x 10 kg | Unit | 01 | 65,00 | 65,00 |
| CM 350 application clamp. | Unit | 02 | 200,00 | 400,00 |
| *Total for equipment and permanent material* | | | | **12.849,00** |
| *etens of bigliographic material* | | | | |
| **Itemisation** | Unit | Quantity | Unit R$ | Total value R$ |
| SAN theme book. | Unit | 02 | 60,00 | 120,00 |
| NDVI theme book. | Unit | 02 | 150,00 | 300,00 |
| Agroecology booklet. | Unit | 30 | 10,00 | 300,00 |
| *Total for bibliographic material* | | | | 720,00 |
| **Total for capital expenditure** | | | | **13.569,00** |

| **Funding costs** | | | | |
|---|---|---|---|---|
| *Pens pe Accessory expenses with imports* | | | | |
| **Itemisation** | Unit | Quantity | Unit R$ | Total value R$ |
| None | Unit | 0 | 0,00 | 0,00 |
| **Total for ancillary import costs** | | | | 0,00 |
| **Funding costs** | | | | |
| *dsseesac per diems* | | | | |
| **Itemisation** | Unit | Quantity | Unit R$ | Total value R$ |
| National per diems | Unit | 20 | 320,00 | 6.400,00 |
| **Funding costs** | | | | |
| *Pe items spent on consumables* | | | | |
| **Itemisation** | Unit | Quantity | Unit R$ | Total value R$ |
| Histological resin - 500 ml - leica | Unit | 02 | 1.367,00 | 2.734,00 |
| Procedure glove 100 pcs, large | Box | 10 | 30,00 | 300,00 |
| Tricorn histokit 60 colours | Kit | 04 | 330,00 | 1.320,00 |
| Coproplus - collector - 360 pcs | Box | 02 | 795,00 | 1.590,00 |
| Graduated glass bottle -100 ml | Bottle | 500 | 8,00 | 4.000,00 |
| **Total for material expenditure c** | **and consumption** | | | **9.944,00** |
| **Funding costs** | | | | |
| *Too bad about the tickets* | | | | |
| **Itemisation** | Unit | Quantity | Unit R$ | Total value R$ |
| Round trip RO land ticket | Unit | 10 | 300,00 | 3.000,00 |
| Return flight tickets | Unit | 02 | 1.000,00 | 2.000,00 |
| **Total for *ticket expenses*** | | | | 5.000,00 |
| **Funding costs** | | | | |
| *Items for third parties (individuals)* | | | | |
| **Itemisation** | Unit | Quantity | Unit R$ | Total value R$ |
| Equipment maintenance services | Unit | 10 | 100,00 | 1.000,00 |
| ***Total for third-party expenses*** | | | | 1.000,00 |
| *Items for third parties (legal entities)* | | | | |

| Itemisation | Unit | Quantity | Unit R$ | Total value R$ |
|---|---|---|---|---|
| Technology expenses | Unit | 2 | 200,00 | 400,00 |
| **Total for *third-party expenses (legal entities)*** | | | | 400,00 |
| **Total Capital and Supplies** | | | | **36.313,00** |

Source: the authors.

Table 6 - Total for grants, capital and funding.

| Activity | Modality | Grant R$ | Quantity | No. of months | Total R$ |
|---|---|---|---|---|---|
| Researcher | EXP -C | 1.100,00 | 01 | 24 | 26.400,00 |
| Student | IEX | 360,00 | 02 | 24 | 17.280,00 |
| **Total Scholarships** | | | | | **43.680,00** |
| **Total Capital + Funding + Scholarships** | | | | | **79.993,00** |

Source: the authors.

## References


BRAZIL. **Decree No. 5.154, of 23 July 2004**. Regulates § 2 of art. 36 and arts. 39 to 41 of Law n°9.394/1996. Available at:<http://www.planalto. gov.br/ccivil_03/_ato2004-2006/2004/decreto/d5154.htm>. Accessed on 1 August 2018.

DANTAS, José Mascena. **Photovoltaic system for isolated communities using ultracapacitors for energy storage**. 2012. 125p. Doctoral Thesis. Postgraduate Programme in Electrical Engineering, Federal University of Ceará.

DE FREITAS, Alair Ferreira; DE FREITAS, Alan Ferreira. Enterprises induced by public policies: reflections from the Programme to Combat Rural Poverty (PCPR) in Minas Gerais. **Interações**, v. 12, n. 2, 2016.

DUBOIS, M.; GILES,K.A.; HAMILTON,J.K.; REBERS,P.A.; SMITH, F. Colorimetric method for determination of sugars and related substances. **Analytical Chemistry**, Washington, v.28, p.350, 1956.

KANO, C.; GENTIL, D.F.O.; CHAVES, F.C.M.; ANTONIO, I.C.; CARDOSO, M.O.; BERNI, R. B. (Ed.) **Memória da reunião técnica olericultura no estado do Amazonas**. Manaus: Embrapa Amazônia Ocidental/UFAM, 2013. 38p.

MACEDO R. L. G. **Basic principles for the sustainable management of agroforestry systems**. Federal University of Lavras, Lavras, MG, 2000.

MORIN, Edgar et al. **The seven knowledges necessary for the education of the future**. São Paulo: Cortez, 2014.

NAKAMURA, EDSON et al. Use of CAD/CAE/CAM tools in the development of electrical and electronic products: advantages and challenges. **T&C Amazônia**, 2003.

NODA, S.N.; MARTINS, A.L.U.; NODA, H.; BRANCO, F.M.C.; MERDONÇA, M.A.F.;

MARDONÇA, M.S.P.; BENJÓ, E.A.; PALHETA, R.A.; SILVA, A.I.C.; VIDAL, J.O. Socio-economic context of family farming in the Amazon floodplains. *In:* NODA, S.N. (Org.) Agricultura familiar na Amazônia das águas. Manaus: EDUA, 2007. p. 23-66.

RIBEIRO, Tina Bimestre Selles. **Rural electrification with individual generation systems using intermittent sources in traditional communities: characterising the obstacles to local development**. 2010. PhD Thesis. USP.

## Editorial board



## Catalogue

Research project funded by CNPq.

Borges, Aurélio Ferreira; Mazutti, Mauri Carlos;
Borges, Maria dos Anjos Cunha Silva
Food security in the Western Amazon: a study applied to project design. - 2018.
65 f.
ISBN 978-613-9-639-66019-3
Book - (Project), 2018.
1. food security. 2. agrobiodiversity. 3.
Black and indigenous ancestry. 4. food production.
Traditional peoples. I. CNPq. II. Title.
CDD - 634.92.

Printed by Books on Demand GmbH, Norderstedt / Germany